Statistical Methods in Longitudinal Research

Volume I

Principles and Structuring Change

This is a volume in
STATISTICAL MODELING AND DECISION SCIENCE

Gerald J. Lieberman and Ingram Olkin, editors
Stanford University, Stanford, California

Statistical Methods in Longitudinal Research

Volume I

Principles and Structuring Change

Edited by

Alexander von Eye

Department of Human Development and Family Studies
The Pennsylvania State University
University Park, Pennsylvania

ACADEMIC PRESS, INC.
Harcourt Brace Jovanovich, Publishers

Boston San Diego New York
London Sydney Tokyo Toronto

This book is printed on acid-free paper. ∞

ACADEMIC PRESS, INC.
1250 Sixth Avenue, San Diego, CA 92101

United Kingdom Edition published by
ACADEMIC PRESS LIMITED
24–28 Oval Road, London NW1 7DX

Library of Congress Cataloging-in-Publication Data

Statistical methods in longitudinal research/edited by Alexander von Eye.
p. cm.—(Statistical modeling and decision science)
Bibliography: p.
Includes index.
Contents: v. 1. Principles and structuring change.—v. 2. Time series and categorical longitudinal data.
ISBN 0-12-724960-5 (v. 1: alk. paper).—ISBN 0-12-724961-3 (v. 2: alk. paper) ISBN 0-12-724962-1 (v. 1; pbk).—ISBN 0-12-724963-X (v. 2; pbk)
1. Social sciences—Research—Statistical methods. 2. Social sciences—Research—Longitudinal studies. I. Eye, Alexander von. II. Series
HA29.S7835 1990
519.5—dc20 89-32965
CIP

Printed in the United States of America
90 91 92 93 9 8 7 6 5 4 3 2 1

Contents of Volume I

Contents of Volume II

Contributors

Numbers in parentheses refer to the pages on which the authors' contributions begin. Numbers in italics indicate contributions to Volume II.

Mark S. Aber (151), *Department of Psychology, The University of Illinois, Champaign, Illinois 61820*

R. Darrell Bock (*289*), *Departments of Education and Behavioral Science, The University of Chicago, Chicago, Illinois 60637*

Jeffrey A. Burr (3), *Department of Sociology, State University of New York at Buffalo, 430 Park Hall, Buffalo, New York 14260*

Clifford C. Clogg (*409*), *Departments of Sociolocy and Statistics, and Population Issues Research Center, The Pennsylvania State University, University Park, Pennsylvania 16802*

Mary Delaney (35), *Department of Human Development and Family Studies, The Pennsylvania State University, University Park, Pennsylvania 16802*

Scott R. Eliason (*409*), *Department of Sociology, The Pennsylvania State University, University Park, Pennsylvania 16802*

Edgar Erdfelder (*471*), *Psychologisches Institut, Universität Bonn, D-5300 Bonn 1, Federal Republic of Germany*

Paul A. Games (81), *Department of Human Development and Family Studies, Pennsylvania State University, University Park, Pennsylvania 16802*

John M. Grego (*409*), *Department of Statistics, University of South Carolina, Columbia, South Carolina 29208*

Randy J. Larsen (*319*), *Department of Psychology, The University of Michigan, Ann Arbor, Michigan 48109–1346*

J. J. McArdle (151), *Department of Psychology, The University of Virginia, Charlottesville, Virginia 22901*

William Meredith (125, *387*), *Department of Psychology, University of California at Berkeley, Berkeley, California 94720*

John R. Nesselroade (3), *Department of Human Development and Family Studies, The Pennsylvania State University, University Park, Pennsylvania 16802*

Trond Petersen (*259*), *Walter A. Haas School of Business, Barrows Hall 350, University of California at Berkeley, Berkeley, California 94720*

David Rindskopf (*443*), *Graduate School and University Center, City University of New York, New York, New York 10036–8099*

Michael J. Rovine (35), *Department of Human Development and Family Studies, The Pennsylvania State University, University Park, Pennsylvania 16802*

Bernhard Schmitz (*351*), *Max-Planck-Institut für Bildungsforschung, Lentzeallee 94, D-1000 Berlin 33, Federal Republic of Germany*

Kathryn A. Szabat (*511*), *Department of Management, LaSalle University, Philadelphia, Pennsylvania 19141*

David Thissen (*289*), *Department of Psychology, University of Kansas, Lawrence, Kansas 66045*

John Tisak (125, *387*), *Department of Psychology, Bowling Green State University, Bowling Green, Ohio, 43402*

Alexander von Eye (*545*), *Department of Human Development and Family Studies, The Pennsylvania State University, University Park, Pennsylvania 16802*

Phillip Wood (225), *Psychology Department, University of Missouri-Columbia, Columbia, Missouri 65211*

Preface

Longitudinal investigations are crucial to the study of development and change. In particular, intraindividual change can be depicted only via repeated observations. To many social scientists, the statistical analysis of repeated measurement data is a challenge because of problems with dependent measures, loss of subjects, availability of computer expertise, or sample size. In many instances, researchers have problems matching substantive questions and statistical methods.

In the methodological literature, considerable progress has been made in the discussion of methods for handling longitudinal data. This progress has remained unnoticed by many, and teaching as well as application of statistical methods has not benefitted from the advances in methodology. In other words, there is a gap between the state of the art of methodology and the application of statistics. It is the main goal of these volumes to narrow this gap.

To illustrate the progress that has been made in the development of statistical methodology, each chapter presents new aspects of methodology or statistics. Here, the attribute "new" applies to all facets of the methods for analysis of longitudinal data. For instance, "new" can mean that methods well-known in certain fields of application are applied to a problem from another field for the first time; it can mean that problems with the application of a software package are identified for the first time; it can mean that a statistical method is further developed to accommodate an extended range of application; it can mean that characteristics of a method are outlined for the first time, so that applicants have a clearer picture of when best to apply this method; it can mean that well-known problems are expressed in terms never used before to express them, so that the researcher has a language available that helps structure problem specification. All chapters in these volumes contribute something new in this sense. None is merely a write-up of a well-known approach.

A second goal of these volumes concerns the application of the new methods. Reasons for the wide gap between the results obtained in the development of statistical methods and their application in empirical research include the lack of computer software and the lack of instructions on how to use available software in applying new methods. This book emphasizes computational statistics. Each chapter that explicitly discusses statistical methods and their application contains instructions and examples of how to use a particular program or identifies available programs. In addition, a companion book by Rovine and von Eye will be published by Academic Press in which examples of program applications are detailed.

The targeted readership of these books includes students of development and change. The books are not oriented toward a particular discipline. Rather, the scope is broad enough to include researchers from all empirical sciences, including the social sciences, economics, biology, or medicine.

The volumes contain 16 chapters, grouped in four sections. Volume I contains Sections 1 and 2, and this volume contains Sections 3 and 4. The first section covers problems of general interest. It begins with a discussion of change processes (Burr and Nesselroade). Here, central terms are explicated, and general problems are specified. The second chapter discusses the problem of missing data, which is almost ubiquitous in longitudinal research in the social sciences (Rovine and Delaney). Approaches to repeated measurement analysis of variance with covariates are discussed in tandem with problems of the application of commercial software in Chapter 3 (Games).

The second section includes chapters on the structuring of change. It begins with a chapter on longitudinal factor analysis (Tisak and Meredith). This approach has only recently been "redetected," drawing renewed attention from both methodologists and applicants. An approach that has changed the perspectives in the domain of longitudinal research is structural equation modeling. The next chapter covers this approach (McArdle and Aber). Rarely applied in present longitudinal research, but nevertheless promising, are methods of scaling. The chapter by Wood introduces readers to this methodology.

This volume begins with the third section, which covers the analysis of time series. It opens with a discussion of event history analysis (Petersen), a method that, thus far, has found most application in sociology but is of great interest to other fields as well. The next chapter treats growth curve analysis (Thissen and Bock), adopting the perspective of mathematical modeling of observed processes. The following chapter, by Larsen, covers spectral analysis. It presents methods of decomposing time series into elementary functions, trigonometric functions in this case. Each of these functions can be interpreted in terms of assumed processes. The fourth chapter in this section discusses time series analysis within the Box and Jenkins framework (Schmitz), including sections on multivariate analysis. The last chapter of this section introduces the reader to methods of segmenting multivariate response curves within the framework of the general linear model (Tisak and Meredith). In recognition of the contribution of Ledyard R Tucker to this methodology, the authors use the term the term "tuckerizing curves."

The fourth section discusses developments in the analysis of repeatedly observed categorical data. It begins with a new approach to formulating longitudinal models for log-linear modeling (Clogg, Eliason, and Grego). The authors adopt the design matrix approach to specifying longitudinal models. Closely related to log-linear modeling is latent class analysis. This approach is covered in the second chapter of this section by Rindskopf, which includes a section on how to use IMSL programs to do parameter estimations. In the third chapter, Erdfelder covers finite mixture distributions. This approach allows one to test assumptions concerning univariate as well as multivariate discrete or continuous distributions. Custom-tailored hypothesis testing in contingency tables can be done with prediction analysis (Szabat), as shown in the fourth chapter, which includes sections describing characteristics of the statistical tests applied to evaluate the model. The final chapter discusses exploratory configural frequency analysis (von Eye). This method allows one to test cellwise whether assumptions concerning longitudinal processes are fulfilled.

I am indebted to a plethora of individuals who supported this work. Viewed from a longitudinal perspective, I would like first to thank J. R. Nesselroade, who encouraged me and helped me to get started. I also would like to thank Academic Press, and in particular Klaus Peters, who was very interested and supportive. Many friends supported this enterprise by reviewing chapters: Constance Jones, Gustav A. Lienert, Jack McArdle, John Nesselroade, Mike Rovine, Holger Wessels, and Phil Wood. I thank them for their efforts and their wisdom. For her invaluable secretarial support, I wish to thank Nancy Cole. I am deeply indebted to the authors who contributed outstanding chapters, often under adverse conditions: One author's computer exploded while he suffered through the drafting process in Hawaii, but he nevertheless submitted a very impressive paper. All authors responded in a very professional way to requests for revisions. Some authors' patience was tried because other authors needed time to complete their chapters. I thank them all very much. Without their efforts, these volumes would not have been possible. Most of all, I would like to thank Donata, Maxine, Valerie, and Julian, who provide the right context and a longitudinal perspective for mutual development in our family.

Alexander von Eye

I General Problems and Issues

Chapter 1 Change Measurement*

JEFFERY A. BURR

Department of Sociology
SUNY–Buffalo
Amherst, New York

JOHN R. NESSELROADE

Department of Human Development and Family Studies
The Pennsylvania State University
University Park, Pennsylvania

Abstract

The representation and measurement of change is a central concern of virtually all scientific disciplines. Studying change directly requires the implementation of longitudinal research designs. From simple difference scores to complicated multivariate models, a variety of methods, procedures, and techniques involving longitudinal design have been developed to assist researchers in the quantification and analysis of change. Many of these tools are identified and examined in this chapter. Obviously, there is not one *best* way to study change. We have tried to identify some important alternatives, many of which are discussed at length in other chapters of this volume.

* This work was partially supported by a grant from the National Institute on Aging (No. T32-AG-00110-02) and by the MacArthur Foundation Research Program on Successful Aging.

1. Introduction

1.1. A Perspective on Change Measurement

The use of longitudinal methods, the ostensible focus of the volume, is intimately tied to the study of change. The representation and measurement of changeable and changing phenomena is very much a concern in nearly all scientific disciplines. Heavenly bodies pass through stages, geological strata shift, social groups form and disband, individuals age, insects metamorphose, and cells divide and multiply. These examples illustrate what we will use as our nominal definition of change throughout this chapter: *change is any variation* in quantity or quality of an entity's attributes.

Social and behavioral scientists, especially those interested in the study of individuals and social collectivities, often study change and change processes. They assume that equilibria, if present at all, are relatively temporary phenomena surrounded by more interesting, yet intermittent periods of change. Wohlwill (1973), for example, proclaimed that "change rather than stability is the norm" (p. 23). Kessler and Greenberg (1981) argued that "whether we are concerned with attitudes, beliefs, or behaviors, whether with individuals, institutions, or whole societies, the central concerns of social scientists revolve around the disruption of patterns of stability and change, and the explanation of how and why change takes place or fails to take place" (p. 1). Many other social and behavioral scientists, both before and after these authors, have expressed similar sentiments regarding the importance of studying change. Moreover, there is a great deal of consensus with regard to the value of longitudinal design and measurement for this purpose. Thus, no longer are observers and sponsors of research enthusiastic about accepting only point-specific or cross-sectional data collection procedures when resources permit multiple time point (repeated measurement) references to human phenomena. Today, the pretest and posttest experimental designs of behavioral scientists, retrospective and prospective panel studies of survey researchers, and time series and trend studies of economists all contribute to the conceptual representation, measurement, and statistical analysis of change. These and other derivative designs are not only important for describing phenomena and analyzing treatment or intervention effects but also for testing hypothesized causal relationships between variables in a system.

Humans, as observers of phenomena, are "hardwired" to respond to variation, variability, novelty, and change. We tend to habituate rapidly to constant stimulation and react to variation. The diffculty for social and behavioral scientists, however, has proven to be in specifying and quantifying the variability and changes of interest. It is to those matters in the context of social and behavioral science, with special emphasis on the study of development by longitudinal research methods, that we will attend throughout this chapter and with which other authors are concerned throughout the volume.

Reliance on individual-level analyses has become an important feature of studying process kinds of changes. Baltes, Reese, and Nesselroade (1977), for example, proposed that developmental research focus on intraindividual change and interindividual differences (and similarities) in intraindividual change patterns. Such a focus rests on the ability to construct individual-level change values and to be able to compare and contrast them in rigorous ways. This emphasis has a significant history, as we will note. Recent attention to modeling growth in terms of latent variables has stimulated important, promising, new developments (see the chapters by Thissen and Bock, McArdle and Aber, and Tisak and Meredith, this volume, and the chapter by Rindskopf, volume II), to which we will turn after examining simple difference scores and their kin.

The reader should bear in mind that there is a long, somewhat troubled (and troubling) history of change measurement in social and behavioral science. Sometimes, for apparently valid reasons, we are told that measuring change is not a very meaningful operation. Seemingly just as often we are told how important it is, especially for developmentalists, to be able to measure change at the level of individual units. The recurrence of debate concerning the measurement of change during the 20th century testifies to two key notions: (1) the topic is of considerable interest to researchers; and (2) proposed solutions have not been universally accepted. Despite some powerful rhetoric questioning the value and validity of change measurement (e.g., Bohrnstedt, 1969; Cronbach and Furby, 1970; Kessler, 1977), interest in how best to represent and measure change continues unabated.

As important breakthroughs occur in conceptualizing developmental processes, modeling humans as living systems, and seeking more complete representations of the sheer complexity of human behavior, of humans in complex, changing contexts, and of developing humans in changing contexts, the need for sophistication in representing and assessing changes becomes ever more urgent. We can ill afford to be satisfied with a situation of unresolved dilemmas and paradoxes (see, e.g., Bereiter, 1963) if we are to have a conceptual and methodological substrate suitable to the requirements of advancing developmental theory and research (Brim and Kagan, 1980).

1.1.1. Measuring Versus Structuring Change. Given that one is committed to change measurement, on what variables or entities should the change scores be produced? The obvious choice is to have measurement variables chosen according to one's research purposes and theory concerns (i.e., manifest variables). Many of the discussions of change measurement and its problems have either implcitly or explicitly reflected this choice. In this chapter we will discuss change measurement at the level of manifest variables. Also, we will attend to the alternative of measuring change on latent variables. The latter alternative will be presented in terms of a multivariate orientation to studying phenomena empirically. Multivariate representations of change have a long history and permit distinctions, such as that made by Harris (1963), between change in level versus change in pattern.

First, we will consider some of the issues and techniques involved in analyzing change in relation to manifest variables (measuring change), including various computations, their strengths, and their weaknesses. We will then examine change measurement in relation to latent variables as they are inferred from interrelationships of manifest variables (structuring change). An imporant link between the two foci is the possibility of applying the concerns of measuring change to linear composite scores constructed to represent the latent variables.

1.1.2. Describing Versus Explaining Change. Science is concerned with both the description and explanation of phenomena. The latter is closely linked to matters of prediction and control. As so many have argued over decades, it is difficult to imagine systematic efforts to develop explanations for phenomena that have not been first adequately described. Obviously, in some circumstances explanations might be stumbled upon before the task of description is completed and, in many cases, attempts at both description and explanation may facilitate improvements in the other in an iterative, recursive process. Cattell (1988), for example, described the process of knowledge generation as an inductive-hypothetico-deductive spiral with description and explanation successively augmenting each other in such a fashion.

In this chapter we will discuss issues that are pertinent to description as well as some that apply to explanation and some that apply to both. In some cases the distinction is somewhat blurred, as, for example, in testing hypotheses in a multivariate analysis of variance (MANOVA) design where the weights for combining dependent variables are determined in the course of the analysis. We try to be explicit concerning description versus explanation in those cases where it might affect the interpretation and use of change concepts.

1.2. Kinds of Data

One of the questions that often arises in discussions of measurement properties concerns whether the issues raised apply only to psychological test scores, an implication being that other kinds of scores are immune to the problems and issues raised. Our own belief is that the matters of reliability and validity we raise are fundamental to measurement, whatever the nature of the measurement process involved. Nunnally (1967), for example, made a useful distinction between *ostensive* attributes, such as length, versus attributes without obvious character, such as psychological traits. Nevertheless, attributes such as length can be measured well or poorly, reliably or unreliably. In a different vein, laboratory manipulations are seldom questioned in regard to quality of measurement criteria. Labouvie (1980), however, made the point that concerns with reliability and validity, usually detailed in consideration of dependent variables, should be extended to independent variables as well.

Thus, for example, in discussing the reliability of difference scores, we will present a formula that indicates how the reliability of a difference measure depends on the separate reliabilities of the two occasions on which that difference is based and on the correlation between the scores at the two occasions of measurement. The levels of reliability may differ systematically, perhaps, but the reasoning applies to differences in measures of speed, growth, and learning just as surely as it applies to differences in psychological test scores.

1.3. Goals of This Chapter

Overestimating the importance of measuring change in the social and behavioral sciences is difficult, especially in the study of human development. Choosing the most useful and appropriate mathematical and statistical representations of change phenomena requires a full understanding of the research design, data characteristics, and the range of methods available to the researcher. It also requires an awareness of the strengths and weaknesses of a given technique. Our central goal in this chapter is to examine briefly some selected literature and ideas about the representation, measurement, and analysis of psychological and sociological change that have emerged in the context of conducting longitudinal research.

One of our foci will be the properties of change scores in relation to calculation and interpretation of these measurements. We will also examine some alternatives to change scores. As a contribution to this literature, we reinforce some suggested extensions to the traditional measures of change, with the hope of spurring others to more fully test and elaborate on these alternatives. It is our task to reach into a stew of many decades and dish up a serviceable meal for researchers needing some sustenance in the pursuit of matters of change in their chosen areas. If, in the process, we manage to give it a little stir or to add a pinch or a dash of something that will further flavor the contents, so much the better.

2. Change Scores and Their Kin

One of the simplest ways to measure change (and probably still the most controversial) is the raw gain or difference score. Whether the interest is intellectual growth, increase (or decrease) in earned income, or change in membership size for an organization or social institution, change scores based on two measurement points are often calculated by the formula

$$d_X = X_2 - X_1, \tag{1}$$

where d_X is the change score, X_2 is the score or value of the variable at time 2, and X_1 is the score or value of the variable at time 1. This representation is simple and apparently sensible, but its use has been fraught with controversy (e.g., Bereiter, 1963; Bohrnstedt, 1969; Cronbach and Furby, 1970).

2.1. "Problems" in the Use of Change Scores

Many of the statistical properties associated with the raw difference score as a change measurement have been zealously attacked by psychometricians and sociometricians. Although it is our intention to remind the reader of some conditions under which difference scores can be useful, we also present a brief discussion of the "pitfalls" of using raw change scores.

One of the most fundamental objections to the use of difference scores stems from validity concerns. Some measures used in social and behavioral science applications are suspected of measuring different attributes at different occasions of measurement (e.g., temperament scales, social indicators). For example, learning, testing effects, and other possible influences can act in the production of second and later measurements, perhaps differentially. Thus, if, for some reason, later measurements reflect different qualitites than do earlier measurements, it jeopardizes the interpretation of difference scores based on those measurements.

Cattell (1966b) gave an example of how distortions due to scaling could impinge on difference scores. Such problems, which are often ignored by users of difference scores, are only resolvable with improvements in the measurement and scaling properties of the scores on the separate occasions of measurement.

Difference scores are also argued to be inherently unreliable. The low reliability of difference scores is seen to be a major drawback to the use of such measures (Bereiter, 1963; Cronbach and Furby, 1970; Furby, 1973; Humphreys, 1961; Lord, 1963; McNemar, 1958; but see also Rogosa, Brandt, and Zimowski, 1982). This argument is based in part on classical test theory and can be presented by using the following postulate:

$$X = X_T + X_E \tag{2}$$

where X is the observed score
X_T is the true score component of X
X_E is the error component of X

The reliability of observed scores reflects the amount of error variance relative to the amount of true variance in the scores. Reliability is defined as the ratio of true variance to observed variance. Since the latter is composed of true and error variances, with no increase in true variance, greater error variance is synonymous with greater unreliability in the measure. An expression for the reliability of difference scores involves both the time 1 and time 2 reliabilities and the correlation between time 1 and time 2 scores:

$$r_{dd} = \frac{s^2_{X_1} r'_{X_1X_1} - 2s_{X_1}s_{X_2}r_{X_1X_2} + s^2_{X_2}r'_{X_2X_2}}{s^2_{X_1} - 2s_{X_1}s_{X_2}r_{X_1X_2} + s^2_{X_2}}, \tag{3}$$

where $s^2_{X_1}$ is the variance of scores at occasion 1
$s^2_{X_2}$ is the variance of scores at occasion 2
$r'_{X_1X_1}$ is the reliability of scores at occasion 1
$r'_{X_2X_2}$ is the reliability of scores at occasion 2
$r_{X_1X_2}$ is the correlation between occasion 1 and occasion 2 scores

This formula is discussed by Lord (1963), for example. When $s^2_{X_1} = s^2_{X_2}$, the formula simplifies to

$$r_{dd} = \frac{\frac{1}{2}(r'_{X_1X_1} + r'_{X_2X_2}) - r_{X_1X_2}}{1 - r_{X_1X_2}} = \frac{r'_{X_1X_1} - 2r_{X_1X_2} + r'_{X_2X_2}}{2(1 - r_{X_1X_2})}, \qquad (4)$$

as presented by Hummel-Rossi and Weinberg (1975), for example.

For a given $r'_{X_1X_1}$ and $r'_{X_2X_2}$, the reliability of the difference scores decreases with increasing $r_{X_1X_2}$. Thus, it is often the case that the difference scores will be less reliable than their time 1 and time 2 components (Hummel-Rossi and Weinberg, 1975). An exception occurs when $r_{X_1X_2}$ tends to be low (or even negative). In that case the reliability of the difference scores can be quite high. However, a low or negative value of $r_{X_1X_2}$ raises the question of whether one is actually measuring the same thing at time 1 and time 2; that is, are the difference scores valid? This is what Bereiter (1963) referred to as the unreliability-invalidity dilemma. We note that in applications involving scores on changeable or labile attributes (e.g., state anxiety), theoretically appropriate low test-retest correlations between first and second measurements can help to produce quite reliable change scores (Nesselroade, Pruchno, and Jacobs, 1986).

Another problem in the use of difference scores to represent change is the relationship between the resulting change score and the initial value (time 1 score) (Benjamin, 1973; Bereiter, 1963; Bohrnstedt, 1969; Cronbach and Furby, 1970; Etaugh and Etaugh, 1973; Furby, 1973; Thorndike, 1942). Here the correlation between change (d_X) and the time 1 value (X_1), $r_{d_XX_1}$, tends to be biased in a negative direction and may not yield a measure of the true relationship between the difference score and the initial value. In essence, a mathematical relationship becomes confounded with the substantive expectation that low initial scores will tend to increase and high initial scores will tend to decrease. Indeed, if the initial scores are extreme enough, they can only change in one direction. Correcting for the bias presents another dilemma in change measurement (Bereiter, 1963).

One of the most frequently cited, troublesome problems associated with using raw gain scores is "regression toward the mean" (Bohrnstedt, 1969; Clarke, Clarke, and Brown, 1959; Cronbach and Furby, 1970, 1970; Hummel-Rossi and Weinberg, 1975; Labouvie, 1982; Lord, 1963; Lytton, Croxen, and Pysh, 1973; Nesselroade, Stigler, and Baltes, 1980; Thorndike, 1942; Vockell and Asher, 1973; and many others). There are at least two issues of concern here: (1) relationships among extreme scores and mean values and (2) errors of measurements. Multiple measures of imperfectly related scores where the first score or value is high at time 1 and lower at time 2 (and vice versa) suggest that unmeasured characteristics may be differentially manifesting effects on such scores, thus leading to erroneous interpretations. For instance, an extreme score (high or low) at time 1 is usually followed by a less extreme score at time 2 (either lower or higher), an expression of regression toward the mean. Obviously, to the extent that it is true, if high scores at time 1 inevitably have lower scores at time 2, and vice versa, then difference scores tend to reflect the regression phenomenon as well, perhaps, as some "real" changes. One needs, however, to look to the nature and pervasiveness of regression effects in evaluating difference scores from this perspective.

A common explanation for regression toward the mean is that, in terms of measurement error, persons with low time 1 scores or values received them in part because of negative error components and the mean error for this subgroup would also be negative. Differencing values across two times would tend to show a mean error score of zero and, thus, a gain in score or value from time 1 to time 2. The reverse would hold for persons with high initial scores and the scores of both groups would demonstrate regression toward the mean.[1]

Thus, as noted, the literature of several decades has contained arguments against using difference scores to represent changes. Despite the reasons for not using difference scores, others (e.g., Tucker, Damarin, and Messick, 1966) have noted that, depending on what aspects of data one wants to emphasize, the use of difference scores may be quite appropriate (see also Maxwell and Howard, 1981; Nesselroade and Cable, 1974).

[1] Another logical problem in the use of change scores concerns differencing raw scores that contain different metrics of measurement. In this case, the researcher is advised to transform the values to a common metric where this makes sense.

2.2. Alternatives to Direct Change Scores

Concern about difference score properties has led to various proposals for "fixing up" raw difference scores to represent change. Proposals include residual scores, "base-free" measures of change, and multiple regression analysis to "predict" time 2 scores from time 1 scores plus other variables of interest.

Residualized change scores are often employed to attenuate some of the problems associated with raw gain scores, as noted earlier. Here the observed score X_2 and a predicted X_2 (e.g., X_2 predicted from X_1 by least squares) can be differenced to yield a value of X_2 residualized with respect to X_1. Thus, the correlation between X_1 and the "new" change score, X_2, is zero and the usual relationship between the initial value and the time 2 value of a score is removed. From this, partial correlations (and sometimes part correlations or unstandardized regression coefficients) can be used to estimate the effect of a control variable (Z) on the relationship between two scores taken at different times. After the partial correlation $r_{X_1X_2 \cdot Z}$ is estimated, a new measure is given that shows correlation between the time 1 and time 2 variables of interest after some important grouping characteristic (Lord, 1963) is held constant. This method assumes a constant metric in the variable of interest, a sufficiently large sample size, and multivariate normal distribution.

These partial correlations also suffer from problems of reliability. Since the observed scores contain error, the correlation between the observed scores is not the same as the correlation between the true scores. To correct for this reliability problem, some authors (Hummel-Rossi and Weinberg, 1975; Tucker, Damarin, and Messick, 1966) suggest dividing the X_1 component correlations by the square root of the reliability of the X_1 terms.

Suppose a researcher has a single item measured at two time points, X at time 1 and Y at time 2, along with a third variable used for grouping, W. A useful formula for correcting the unreliability in the time 1 measure, X, is

$$r_{WY.X_T} = \frac{r_{WY}r_{XX} - r_{WX}r_{YX}}{(r_{XX} - r_{WX}^2)(r_{XX} - r_{YX}^2)}. \tag{5}$$

A similar correction can be made for the time 2 score Y, but this is deemed not as critical. For a comparison of differences scores, residual scores, and correlational approaches to estimating changes, see Tucker, Damarin, and Messick (1966).

Another alternative to raw gain scores is a standardized or relative measure of change (Tornqvist, Vartia, and Vartia, 1985) that has symmetric, additive, and normalized properties and can be employed when change is estimated for variables with different metrics. This method assumes a ratio scale of measurement and positive integers for the variable of interest. This alternative method of computing change scores relies on the mathematical properties of the natural log to eliminate between-occasion differences in metric scores. The formula for this measure is

$$\log_e(X_2/X_1) = \log_e X_2 - \log_e X_1 = \Delta \log_e X. \tag{6}$$

Defining the measure explicitly in this way avoids the issue of which value (time 1 or time 2) to use as a base to calculate relative change.

Orwin (1981) demonstrated that "base-free" change scores can be problematic if certain assumptions about the baseline measures are violated. Specifically, Orwin suggested that if "extraneous effect factors" are present and are not equivalent across observations points, the baseline measures will change from time 1 to time 2. This will result in spurious effects on the estimates of change. Such change scores give misleading results and are impacted even further when "true" treatment effects are operating. This concern about spuriousness in change score estimates is related to the discussion by Liker, Augustyniak, and Duncan (1985; see Section 2.3).

2.3. Closing the Circle on Change Scores

The enormous amount of discussion and negative critical comment on change scores during the 1950s, 1960s, and 1970s created an atmosphere that led many researchers to shun the use of these measures. However, the psychometric and sociometric literature is gaining momentum toward a reaffirmation of the value of change measurement (see, for example, Liker, Augustyniak, and Duncan, 1985; Maxwell and Howard, 1981; Nesselroade and Bartsch, 1977; Nesselroade and Cable, 1974; Nesselroade, Stigler, and Baltes, 1980; Rogosa, Brandt, and Zimowski, 1982; Rogosa and Willett, 1985).

Nesselroade and Bartsch (1977) and Nesselroade and Cable (1974) argued for the meaningfulness of difference scores in situations involving measurement of state attributes that reflect relatively short-term changes. Within the framework of the common-factor model and two occasions of measurement, they demonstrated that difference scores follow straightforward rules regarding the essential nature of statelike versus traitlike variables. Nesselroade, Stigler, and Baltes (1980) showed that when the researcher moves beyond the usual two occasion measurement scheme to the multiple measurement scheme (three or more) that at least three possibilities exist: (1) $r_{X_1X_i}$ decreases as i increases, (2) $r_{X_1X_i}$ increases as i increases, and (3) $r_{X_1X_i}$ remains stable as i increases (1980, p. 636). There is regression toward the mean in all three cases, and, in case (1), regression continues past the second sequence. However, in case (2), after sequence 3, egression occurs, and in case (3), after the third measurement point, stability is witnessed. One of the major contributions of this work is that it alerts social and behavioral researchers to the fact that more than two points of measurement are logical ways to proceed with research on change in humans, and that change scores can be useful under these multipoint data collection conditions. We return to this point later.

Others have taken up the call in defense of change scores (Maxwell and Howard, 1981; Overall and Woodward, 1975; Games, this volume). These authors promote the usefulness of change scores in randomized pretest-posttest designs in the presence of response-shift bias and multivariate measurements. In the pretest-posttest design where one-way univariate ANOVA on the posttest scores is usually recommended, Maxwell and Howard (1981, p. 749) show that the ANOVA approach to the difference between posttest and pretest yields the same results. In addition, if the time 1 and time 2 measures are correlated, the difference approach may be more powerful, although the analysis of covariance is often preferred over either of these approaches.

When there is response contamination (response-shift bias), Maxwell and Howard (1981, pp. 750–751) claim that the use of change scores is the *only* method that will yield meaningful results. The elimination or attenuation of contamination is accomplished by applying a retrospective pretest at the posttest time point to ensure that the respondent is rating his pretest and posttest on the same "standard" of evaluation. In a multivariate pretest-posttest design, change scores are also recommended. In situations where the number of variables is large relative to the sample size, the larger degrees of freedom associated with the MANOVA change scores may provide more statistical power than the MANOVA.

Liker, Augustyniak, and Duncan (1985) have made convincing arguments for the use of first-difference equations as opposed to more conventional two-wave, two-variable models under certain restrictive conditions. The authors begin their argument with the general change equation (p. 83)

$$\Delta Y = \Delta B_0 + \Delta B_1 X + \Delta B_2 Z + \Delta e, \tag{7}$$

where Δ equals simple change from time 1 to time 2, and the subscripts for each case, *i,* are assumed. If the Z variables are unmeasured and unchanging, yet important contributors to the explanation of Y, then differencing the equation removes the contaminating effect of Z. This results reduces the regression to ΔY on ΔX.

These authors also argue that given measurement error in X, a differencing equation process will negate the deleterious effects of measurement error if there is perfect autocorrelation ($r_{e_1e_2} = 0$). In addition (given that perfect autocorrelation is rare, especially in survey data), differencing positive, yet imperfect, autocorrelation will yield a smaller bias in the effect of measurement error. In a less convincing manner, the authors also argue that in some empirical situations panel data give more reliable knowledge of changes in X over time than in the level of X at a given time. This argument is based on "memory error" in retrospective surveys and is not fully fleshed out in the paper.

Liker, Augustyniak, and Duncan (1985, pp. 85–94) proceed with a noteworthy discussion of the algebraic relationship between differencing equations and two-wave, two-variable models (Duncan, 1969; Lazarsfeld, 1948; see Kessler and Greenberg, 1981, for a full decomposition of the mathematical relationships). The two-wave, two-variable (2W2V) model originally believed to be a panacea for issues concerning the analysis of cause and effect is

$$X_2 = a_0 + B_1 X_1 + B_2 Y_1 + B_3 Y_2 + e, \tag{8}$$

$$Y_2 = a_1 + B_1 Y_1 + B_2 X_1 + B_3 X_2 + e. \tag{9}$$

This model is known to be underidentified and will not yield unique solutions without algebraic manipulation. Some have argued that to fully exploit panel data without using change scores and to account for the negative bias associated with a correlation between initial value of the dependent variable and change in that variable over time, the initial value of the dependent variable should be included on the right side of the regression equation ("cross-lagged panel model"; e.g., Bohrnstedt, 1969). Liker and co-workers argue that this is inappropriate for estimating stability effects.

First, often there is no theoretical justification for assuming that the initial value of the dependent variable should explain the time 2 value, and this results in a loss of parsimony (it is true that in many empirical cases time 1 values of a variable *predict* time 2 values, but this does not necessarily help *explain* the empirical change relationship). Second, the error structure of the time 1 values of the dependent variable will most likely be correlated with the time 1 values of the independent variable(s), creating problems of bias and violation of basic regression assumptions. In sum, initial values of the dependent variables should not be automatically included on the right side of the regression equation when trying to avoid regression toward the mean. When it is necessary to estimate this effect, instrumental variables should be employed to avoid statistical problems introduced under these conditions.

Liker, Augustyniak, and Duncan (1985) were not the first to be concerned about the usefulness of cross-lagged correlations for studying change and causality. Several statisticians have noted that the basic 2W2V model and its extensions to large systems (in terms of more variables and more waves of data) have notable drawbacks (see, for instance, Borhnstedt, 1969; Duncan, 1969; Goldberger, 1971; Heise, 1970). Perhaps no one has been more critical than Rogosa (1980). Rogosa states that the 2W2V model is "useless" not only for the evaluation of causality through the examination of the difference between a set of cross-lagged correlations but also for examining possible spuriousness among the variables (Kenny, 1975).

Rogosa states that regardless of whether the approach is with structural regression models, continuous-time models, or multiple time-series models, the underlying assumptions of stationarity, synchronicity, and homogeneous stability are rarely considered and often violated. The violation of these assumptions, which Rogosa considers to be frequent and severe, wreaks havoc on attempts to determine causal structures. The substantive researcher is also warned that specification and measurement error contribute to the lack of robustness in the 2W2V model's ability to establish causal relationships. Rogosa finally states the same sentiment that Duncan and others have been making for years—that is, no methodological procedure will substitute for the fundamental necessity of integrating theory and research design.

3. Alternatives to Change Scores

The foregoing discussion lists several methods for generating change scores along with critiques on the strengths and weaknesses of each. In addition, we present evidence from the literature that shows that under certain restrictions, and for specific research questions and designs, change measurement should not be "feared" or "loathed" as an intractable, statistical wasteland.

As researchers concerned with the identification and study of change patterns, we have to believe that the basic idea of measuring change is sound. The question is how best to represent and measure changes. Obviously, this depends to a large extent on the kind of change phenomenon that one is trying to study. Although instances have been noted in which simple change scores are seemingly appropriate to use, we now want to turn to some alternatives that promise to be valuable for developmentalists in exploiting longitudinal data.

In this section we provide the reader with methodological alternatives used in the analysis of change as well as related issues in causal modeling. We also suggest how Cattell's "databox" formulation (1966a) can be 'sliced" or partitioned differently to lead toward new considerations on the measurement of changes. Before turning to the data box, however, we briefly examine some suggestions by authors who have wrestled with the problems of change measurement in a developmental or other process context.

Labouvie (1982) suggested that we consider using change in variances instead of changes in means as a complementary approach to the analysis of change. Most treatments of change rely on patterns of autocorrelations to the neglect of patterns of autocovariances (Labouvie, 1982, p. 251), in large part due to the statistical convenience of assuming homogeneity of variance. The point is well taken that assuming constant variance across time in regard to intraindividual change and aging is not necessarily realistic. The deviation change score is defined as

$$X_2 - X_1 = W_i + Q_i - W_i = Q_i, \tag{10}$$

where W_i is the deviation from the average for person i and Q_i is the deviation from the average change for person i. The standard change score is defined as

$$\begin{aligned} \frac{X_2}{s_2} - \frac{X_1}{s_1} = Z_2 - Z_1 &= \left(\frac{W_i + Q_i}{s_2} - \frac{W_i}{S_1}\right) \\ &= \left(\frac{1}{s_2} - \frac{1}{s_1}\right) W_i + \frac{Q_i}{s_2}, \end{aligned} \tag{11}$$

where s is the standard deviation (Labouvie, 1982, p. 253).

If homogeneity of variance does not represent reality, then the standard change score is not a "straightforward" linear transformation of the raw gain score: the standard change score is a linear combination of the initial differences and the deviation score (1982, p. 253). Thus, the two types of change measures will yield different results, and the standard change score is biased with respect to r_{WQ}. In general, Labouvie makes a case for the more general applicability of deviation scores and goes on to demonstrate how such an approach can aid in the prediction of change and in the modeling of causal processes.

3.1. *Qualitative Change Models*

There are two broad classes of change models that statisticians and other researchers have at their disposal when examining psychological, sociological, and economic dynamics: *qualitative* and *quantitative* change processes (Hannan and Tuma, 1979; Tuma and Hannan, 1984). The choice of which type of model to employ depends on the nature of the data and the assumptions concerning the relative importance of the type of change involved. For instance, the researcher must determine whether change from time t to time $t + 1$ is more likely to be continuous or discrete in the "real" world. Both theory and past empirical results guide the choice of models. Data collection techniques also play a large role in the choice of methods. Panel studies, whether of the repeated–measurement variety or for event histories of varying degrees of completeness, represent the most common type of multipoint, qualitative data in the social and behavioral sciences.

Up to this point, we have surveyed and discussed the quantitative change literature as related to the assumptions, strengths, and weaknesses inherent in these approaches. In this section, we review some of the various methods for measuring change in qualitative outcomes (discrete-state). Qualitative approaches are defined by movement from state i to state j often, but certainly not exclusively, in the dichotomous case (e.g., anxious to relaxed, married to not married, employed to unemployed, etc.). Several of the methods discussed here can handle multinomial responses in change measurement.

Social scientists have long been interested in using panel models to explain change and to evaluate causal relationships (Lazarsfeld, 1940; Lazarsfeld, Berelson, Gaudet 1944). Log-linear models developed and refined by Goodman (1972, 1978), Bishop, Fienberg, and Holland (1975), Clogg (1981, 1982), and others have been extended for the treatment of change (see also Clogg et al., this volume). A prime example of the combined interests of measuring change and causal relationships can be found in intergenerational and intragenerational mobility table analysis, which is often, but not always, concerned with occupational movement (for practical applications see Hauser et al., 1975a; 1975b). The minimum data required for such analyses is a panel with two time points, which records origins and destinations of the respondent in square matrix form on some important characteristic (occupation of father and occupation of son, for example).

The standard log-linear model, which can be used to analyze a 2×2 table of association is

$$\log(F_{ij}) = a_0 + a_{1i} + a_{2j} + b_{ij}, \tag{12}$$

where a_0 represents the grand mean

a_{1i} is the row effect

a_{2j} is the column effect

b_{ij} is the interaction effect

In this formulation, the sum of the row and column parameters must equal zero, and the sum of the interaction effects for the rows and the sum of the interaction effects for the columns must equal zero.

This technique has three major weaknesses: two measurement issues and one sample issue. First, if the variable of interest (i.e., the dependent variable) is better measured as a continuous process, then this approach may not adequately represent such a dynamic process. Also, when independent variables (i.e., covariates) are introduced into log-linear models, they must be inherently categorical or, if intervally measured, categorized. This may reduce the amount of useful information found in the data. A second problem has to do with data collection in panel studies. Often the choice of lags between time 1 and time 2 is arbitrary and does not necessarily represent the true lag (or distance) between two states. This is often the case when the researcher finds himself or herself analyzing secondary data. Third, if the sample size is small, the presence of many empty cells leads to difficulty in estimating parameters.

Two developments in this area deserve brief mention. First, Clogg et al. (1989, Volume II) have proposed a set of models stemming from the general log-linear model to examine change with repeated measures in panel data representing two to five waves of measurement. These authors develop change models and provide empirical examples for discrete outcomes related to (1) the 2×2 contingency table, (2) the $i \times i$ table (more than two outcomes, including special types of contingency tables, i.e., marginal homogeneity, symmetry, quasi-symmetry, and quasi-independence), (3) the $2 \times 2 \times 2$ table for repeated measures, (4) the $i \times i \times i$ table, and (5) models of change with fixed covariates (extensions to the standard logit regression model). Clogg et al.'s description (this volume) of these log-linear techniques for measuring change as provided here will undoubtedly lead to further evaluation and research.

Another variation on the general log-linear model approach, configural frequency analysis (CFA: Lienert and von Eye, 1987), has been explicitly used to model longitudinal change assessed at the individual level (von Eye and Nesselroade, 1989). CFA is concerned with the examination of "types," where the observed cell frequency (within a contingency table) occurs more often than is expected by chance alone, and "antitypes," where the observed cell frequency occurs less often than is expected by chance alone. In the simplest case this technique provides the statistician and researcher with the ability to estimate whether the cross-tabulated frequency of two attributes or levels of two variables are statistically related (i.e., greater or less than expected by chance). The authors suggest three different ways in which CFA can be applied to longitudinal data. The three methods of change analysis with CFA presented in this work are (1) the analysis of trends, (2) the analysis of the timing of change, and (3) the analysis of trends and the timing of change employing orthogonal polynomials (see von Eye in this volume for further treatments of this method for change measurement).

Regression techniques for analyzing dependent categorical or dummy variables have been part of the statisticians bag of tricks for many years (see Goldberger, 1964, for early treatments of linear probability models with discrete outcomes). The estimation of change across two or more states is accomplished most often within the framework of either logit or probit regression models. Of course, other dependent variable distribution assumptions are possible, such as the tobit (see Tobin, 1958).

These regression techniques allow the researcher to estimate the nonlinear probability of moving from one state to another (e.g., from married to not married). One of the advantages of these approaches is that they give the statistician and researcher the ability to employ both discrete and continuous predictor and explanatory variables in the model, which some other techniques do not allow (see Clogg et al. (1989), V.II). One of the drawbacks with these procedures, however, is that they lack the quality of being able to address adequately the issue of count, sequence, and timing in moves or changes (see, e.g., Allison, 1982;

Featherman and Petersen, 1986; Heckman and Singer, 1984; Tuma and Hannan, 1984). Choosing between logit and probit regression techniques depends on the assumed response distributions. A general logit regression model that considers whether a respondent has moved or changed states from time t to time $t+1$ (e.g., changed states from one interview to another) can be represented by the formula

$$\log[P_i(1 - P_1)] = \sum b_k X_{ik}, \tag{13}$$

where P represents the probability of being in cell i at time 2

b represents the regression coefficients

X_{ik} represents the matrix of individual characteristics that condition the probability of movement from time 1 (wave 1) to time 2 (wave 2)

Another advantage of this technique for measuring discrete change is the ability to examine multinomial, discrete dependent variables (see Hanushek and Jackson, 1977). With the advances in statistical software programs, social and behavioral scientists can now more easily employ nonlinear regression to estimate discrete changes in state with both interval (ratio) and nominal (ordinal) independent variables.

A final class of methods that is gaining popularity among psychologists, sociologists, and economists is event history analysis, often analyzed with hazards modeling and employing both parametric and nonparametric estimation techniques. Important technical discussions, empirical examples, and contributions to the method are presented by Allison (1982), Featherman and Petersen (1986), Heckman and Singer (1984), Tuma and Hannan (1984), and Petersen, this volume. The importance of this method can be explicated by referring to its strengths over other types of qualitative change analysis. For instance, with hazards modeling the researcher can estimate models that are not restricted and biased by the use of discrete and often arbitrary lags because a continuous-time, discrete-state transition is assumed (Coleman, 1964, 1968). In addition, covariates of both the quantitative and qualitative variety can be used without relying on a data reduction strategy related to the discretization of interval data. Finally, in its full form, these models can estimate not only the type of change but also the sequence and timing of changes.

The minimum requirement for event history analysis is event count data, where the number of changes in state is measured. The next level is event sequence data, where the count and order of the changes are known. Finally, full event history data are the most useful where count, order, and timing are all known. In other words, event history data applied to "hazards" models allow the statistician and substantive researcher to take full advantage of the nature of the many longitudinal data sets currently available (e.g., the Panel Study of Income Dynamics, Survey of Income and Program Participation, National Longitudinal Surveys, Seattle and Denver Income Maintenance Experiments, etc.). Continuous-time, discrete-state Markovian models are applied in the hazards modeling approach (Coleman, 1964, 1968) and a general formula for estimating the continuous rate of change in discrete outcome without covariates is

$$h_{ij}(t) = \exp[a_{ij}(t)], \tag{14}$$

where $i, j = 1, \ldots, n$ is the index of dependent variable
t refers to time since start of spell
$h_{ij}(t)$ is the risk of moving from state i to state j at duration t
$a_{ij}(t)$ are parameters to be estimated

An extension of this class of hazards models includes the introduction of covariates constructed to estimate the effects of other variables on the rate of transition from state i to state j. A general formula is

$$h_{ij}(t) = \exp[a_{ij}(t)] + \mathbf{B}_{ij}\mathbf{X}, \tag{15}$$

where i, j, t, $h_{ij}(t)$, a_{ij} and variables defined earlier
$\mathbf{X}$ is a set of covariates defined earlier and measured at the start of the spell
$\mathbf{B}_{ij}$ is a vector of parameters to be estimated and associated with $\mathbf{X}$

Despite the promise of this type of change analysis, it has several problems.

The first type of problem is called *unobserved heterogeneity*. The strategies previously described assume that the variables included in the estimating equations capture the relevant impacts on the underlying hazard rate; if additional sources of heterogeneity are omitted from the hazard model, the estimate of specific hazard coefficients may be affected, and, perhaps more problematic, the estimates of sampling variance can be too low. Second, *time dependence* is known to bias estimates of transition rates. Transition rates may vary over time because factors causing or influencing the rates of change may vary over time. The three most common factors related to time dependence include, age, duration in a state, and experience. A third problem is *censoring*. Hazard rates can be successfully generated only for episodes with known start times; when this expectation is absent, statisticians refer to the condition as left-censoring. The problems associated with right-censoring are of less concern in the estimation of these models because all respondents can be assumed to have ended their time in a given state at the same time (end of survey or data collection time point). Despite these problematic issues, the implementation of event history analysis with hazards models appears to have a bright future as a part of the social and behavioral scientist's change measurement strategies.

Our goal for the first part of this chapter has been to review the relevant strengths and weaknesses of well-known change measurement and analysis techniques of both the quantitative and qualitative varieties. The second goal of this chapter is to extend this literture by providing the reader with a discussion of some developments in the analysis of change in latent variables instead of manifest representations. In the next section, we show how the data box can be partitioned to analyze change in latent variable patterns using what are known as T- and P-techniques (Cattell, 1966a).

4. Structuring Change

As noted earlier, an alternative to the use of difference scores to represent change is the reliance on multitime point measurements to construct curves or trajectories of group and of individual change functions. This emphasis has the major advantages of avoiding some of the problems associated with the use of difference scores per se and provides for a more thorough representation of change than is given by one or even a few difference scores, except in those cases where the nature of the change is well captured by differencing methods.

In the introduction, the possibility of focusing on change at the level of latent, rather than manifest, variables was mentioned. There are two aspects to be considered here: (1) change in the patterning of relationships among manifest variables as a basis for inferring change of a more or less qualitative nature; and (2) quantitative changes in latent variables. Qualitative changes are illustrated by, for example, differentiation of performance as manifested in changes in factor patterns (Baltes and Nesselroade, 1970; Burt, 1954; Garrett, 1946) and quantitative changes by shifts in factor scores. These distinctions were reviewed by Nesselroade (1970, 1977) in the context of research in adult development and aging. Suffice it to say that they involve using the common-factor model to distinguish between different kinds of change in repeated, multivariate measures. Here we will focus primarily on two approaches to change in multivariate data: those identified by Cattell (1963) as P-technique and T-technique.

4.1. The Data Box, P-Technique, and T-Technique

Cattell (1952) originally introduced the data box concept as a parallelepiped, the axes of which were defined by persons, variables, and occasions. A version of the data box is presented as Figure 1.1. Within that framework, he enunciated a set of six possible covariation techniques, each involving the covariation of values either across the rows or columns of the three two-dimensional arrays definable within the three-dimensional data box. For example, R-technique involves the covariation between variables computed over individuals, each of whom

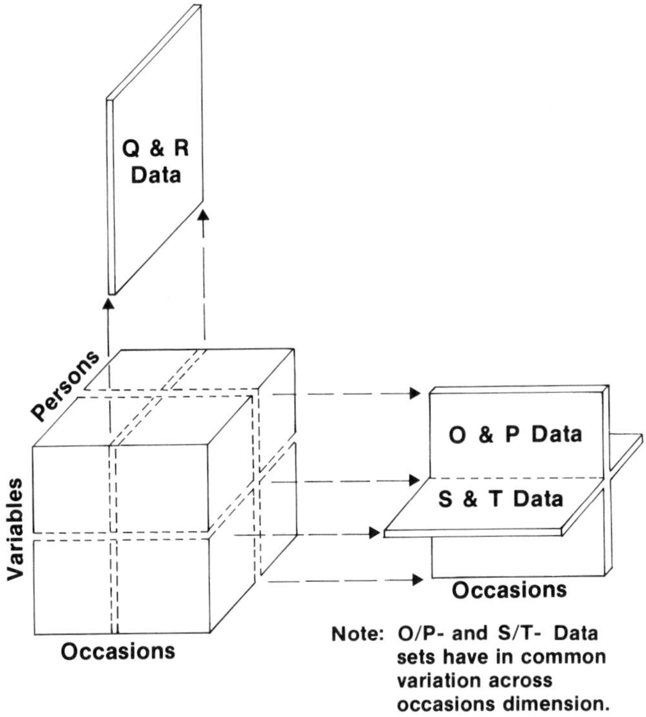

FIGURE 1.1. The data box (after Cattell, 1952). (Reprinted with permission of Springer-Verlag from "Adult Personality development: Issues in assessing constancy and change," in A. I. Rabin, R. A. Zucker, R. A. Emmons, and S. Frank (Eds.), *Studying persons and lives,* New York: Springer, 1990.)

has been measured on the variable set on only one occasion. Q-technique involves the same "slice" of data, but, for analysis, the individuals are covaried across the variables. Subsequent versions of the data box or basic data relations matrix (BDRM) have included up to 10 dimensions (Cattell, 1966a, 1988).

One of the key implications of the data box heuristic is the idea that empirical data typically underrepresent one or more of the modes of classification. This underrepresentation is a kind of selection, and it leads to selection effects that can distort the inferences drawn from empirical studies (Nesselroade, 1983, 1986, 1988a). We cannot go into detail here with respect to the modes of the data box, but it does seem worthwhile to mention the occasions mode with regard to the effects of selection.

The essential argument is as follows. The validity of inferences about relationships based on empirical information rests on the quality of that information relative to the universe to which the inferences are to be made. Biased samples of information do not yield inferences as valid as unbiased ones do. The inferences that developmentalists and others interested in change (and for that matter stability) make depend on the quality of information relevant to the occasions mode of the data box. Data that represent the occasions mode well provide more valid inferences about change (and stability) than do data that sparsely represent the occasions mode. For example, two occasions of measurement will yield a difference or change score, but it may not adequately represent the subtleties of a change function, as shown in Figure 1.2. Multiple occasions of measurement can provide a more accurate picture

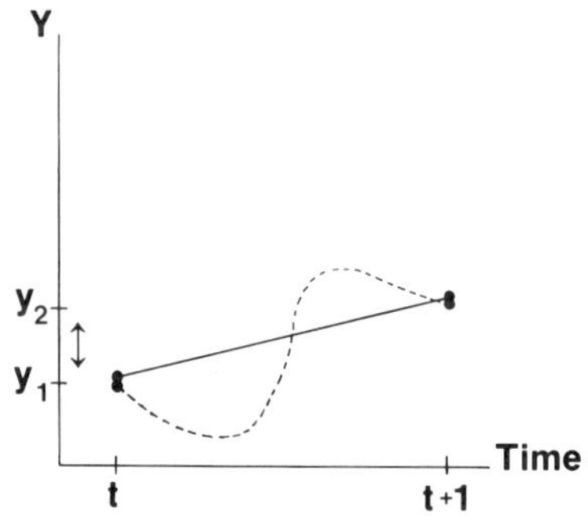

FIGURE 1.2. Illustration of the inability of the selection of two occasions defining a change score to capture the nature of a developmental trajectory. (Reprinted from Nesselroade, J. R., and Ford, D. H. (1987). Methodological considerations in modeling living systems. In M. E. Ford and D. H. Ford (Eds.), *Human as self-constructing living systems*. Hillsdale, N. J.: Lawrence Erlbaum Associates.)

of the nature of change, but obtaining them can require vastly greater resource expenditures. Thus, there is a set of design and measurement issues that need to be considered in light of what is known about the phenomenon in question, the researcher's goals, resources available, etc. Two occasions of measurement are the minimum number from which to compute change scores, and, as Rogosa, Brandt, and Zimowski (1982) pointed out, two occasions *may* not be much of an improvement over only one.

Good descriptive work early in the investigation of a given phenomenon may well require quite intensive measurement emphases. Similarly, explanatory work with change functions may require more or less intensive measurement designs if the richness of the function to be accounted for is to receive proper representation.

P-technique was introduced (Cattell, Cattell, and Rhymer, 1947) as a way to capitalize on repeated measurement (longitudinal data) to investigate intraindividual variability. In P-technique, one person's responses or measurements on many variables are intercorrelated across multiple occasions and factor analyzed. This kind of analysis provides information concerning patterns of covariation among variables across time for a particular individual. These patterns of change are of interest in their own right for studying the nature of change both within individuals and then between individuals. The patterns can also be used as a basis for constructing composite scores (e.g., factor score estimates) to describe changes in level over the period of observation. Discussion concerning the appropriateness of intercorrelating and factoring repeated multivariate observations of the single case are available (e.g., Cattell, 1963; Holtzman, 1963; McArdle, 1982; Molenaar, 1985). Molenaar (1985) has provided the clearest explanation to date of issues pertaining to modeling the dynamics of repeated measurements of individual cases by means of factor analysis.

By the very nature of the data and the operations performed on them, P-technique factors are free of both interindividual differences and intraindividual stability. Rather, they represent the covariation pattern of changes within the individual over the time interval of the observation sequence. The rationale is that covariation among the measurements on different variables rests on two attributes of the data. First, a necessary condition for covariation is that both variables manifest variation. Unless both variables involved in the construction of a given covariance or correlation show variability over occasions, there can be no correlation between them. Second, even if both variables manifest variability across occasions, they do not correlate or covary unless they reflect some common change process, either directly or through the influence of a third variable. The purpose of the factor analysis is to provide a parsimonious representation of the covariation pattern over the set of variables and occasions involved. Bereiter (1963), for example, identified such multivariate single-case studies as "the logical way to study change."

P-technique illustrates another important aspect of change on which we will comment. That has to do with the selection argument introduced earlier, namely, two occasions of measurement define one interval and one of anything is generally a very sparse basis on which to rest generalizations. With regard to time intervals, this seems especially true when one is monitoring an ongoing process, such as individual development. One of the promising current emphases is the development of latent growth curve analysis (see McArdle, this volume; McArdle, 1988; Tisak and Meredith, this volume), which focuses on change as a process rather than as something that is adequately captured with difference scores or some derivative based on two occasions of measurement.

Substantive areas in which P-technique has been applied include mood, emotion, personality traits and states, human abilities, and motivation (see Jones and Nesselroade, 1989, for review). Findings from these studies do show that occasion-to-occasion intraindividual variability (change) is coherent, and, in many cases, the patterning of variability across observed variables is replicable across individuals.

Elsewhere (Nesselroade, 1988b), the argument was made that the differences extant among individuals at one point in time represent both stable interindividual differences and temporary interindividual differences that reflect the fact that at a given time different people happen to be at different levels on change dimensions. A complete understanding of individuality, whether the focus is stability or change, must take this into account. Thus, for a given domain of measurement, it is imperative that something is known about the stability versus changeability of the attributes. It may not be necessary to go to the extremes of intensive measurement represented by P-technique, but certainly some significant sampling of information vis-à-vis the occasions mode of the data box is called for if change (and stability) are to become properly characterized and explored.

4.1.1. T-Technique. T-technique involves data obtained by repeated measurements of a sample of persons (or other experimental units) on a single variable. For example, the recall scores of 100 persons on 10 trials at memorizing a list of nonsense syllables would constitute a 100×10 "slice" of T-technique data. Like P-technique, T-technique involves successive occasions of measurement, seemingly a sine qua non for studying process and change. Unlike P-technique, T-technique involves a sampling of individuals rather than only one person.

Although the uses of T-technique approaches (whether or not they are called that) have been in the literature for many decades, work over the past two decades has begun to show great promise of payoff for students of change processes such as growth, learning, maturation, social processes, etc. For example, after the initial development of methods for the general analysis of learning curves (Rao, 1958; Tucker, 1958, 1966), work on latent growth curve modeling (Meredith and Tisak, 1984, McArdle, 1988; see also McArdle and Aber, and Tisak and Meredith, this volume and Thissen, Volume II) has produced some sophisticated ways to model T-technique data. With the availability of these methds, researchers interested in change now have a wide range of possibilities for structuring and measuring change as manifested in T-technique data.

5. Conclusion

There can be no serious disagreement with the sentiment that change, at least in concept, is an important focus of scientific research. Behavioral and social scientists have tended to be much more self-conscious about actually measuring change than have researchers in other disciplines. Although (largely from a measurement perspective) there are good reasons for some reticence, much of what we want to study as developmentalists is still rather inextricably bound up with conceptions of change, longitudinal data, and, therefore, change measurement. In this chapter we have examined various possibilities for assessing change, some of which amount to variations on the calculation of difference scores based on two measurements of a phenomenon, and others are representations of processes of change and extend, at least implicitly, over many measurements of the phenomenon in question.

We have tried to point out some of the disadvantages as well as advantages of alternative representations of change. In some cases, this has resulted in implicit warnings against using those techniques if robustness under violaton of assumptions is lacking. In any case, the most fundamental message to be conveyed can be couched in terms of generalizing. Committing oneself to a particular form of change representation should involve explicit consideration of the important paths of generalization from those changes to more general inferences concerning the pertinent phenomena. For example, had other information been derived from the data box, would the generalizations based on it converge in appropriate ways with those being made concerning changes based on the data actualy in hand? We are, after all, in the business of drawing inferences from data in hand to the larger, not-ever-to-be-fully-explored data box of nature. Concerns with time and change impel us to gather many of those data within the framework of longitudinal research designs. The means for exploiting such data lie in the evolution of the statistical tools geared to the representation and measurement of change.

References

Allison, P. D. (1982). Discrete-time methods for the analysis of event histories. In S. Leinhardt (Ed.), *Sociological methodology, 1982* (pp. 61–98). San Francisco, CA: Jossey-Bass.

Baltes, P. B., & Nesselroade, J. R. (1970). "Multivariate longitudinal and cross-sectional sequences for analyzing ontogenetic and generational change: A methodological note," *Developmental Psychology*. **2,** 163–168.

Baltes, P. B., Reese, H. W., and Nesselroade, J. R. (1977). *Life-span developmental psychology: Introduction to research methods.* Monterrey, CA: Brooks/Cole.

Benjamin, L. S. (1973). "Remarks on behalf of change scores and associated correlational statistics: A response to the Etaughs," *Developmental Psychology*. **8,** 180–183.

Bereiter, C. (1963). Some persisting dilemmas in the measurement of change. In C. W. Harris (Ed.), *Problems in measuring change.* Madison, WI: University of Wisconsin Press.

Bishop, Y. M., Fienberg, S. E., and Holland, P. W. (1975). *Discrete multivariate analysis: Theory and practice.* Cambridge, MIT Press.

Bohrnstedt, G. W. (1969). Observations on the measurement of change. In E. F. Borgatta (Ed.), *Sociological methodology* 1969 (pp. 113–136). San Francisco: Jossey-Bass.

Brim, O. G., Jr., and Kagan, J. A. (1980). Constancy and change: A view of the issues. In O. G. Brim, Jr., and J. A. Kagan (Eds.), *Constancy and change in human development* (pp. 1–25). Cambridge: Harvard University Press.

Burt, C. (1954). "The differentiation of intellectual abilities," *British Journal of Educational Psychology*. **24,** 76–90.

Cattell, R. B. (1952). "The three basic factor-analytic research designs—their interrelations and derivatives," *Psychological Bulletin*. **49,** 499–520.

Cattell, R. B. (1963). The structuring of change by P-technique and incremental R-technique. In C. W. Harris (Ed.), *Problems in measuring change.* Madison, WI: University of Wisconsin Press.

Cattell, R. B. (1966a). The data box: Its ordering of total resources in terms of possible relational systems. In R. B. Cattell (Ed.), *Handbook of multivariate experimental psychology.* Chicago: Rand McNally.

Cattell, R. B. (1966b). Patterns of change: Measurement in relation to state-dimension, trait change, lability, and process concepts. In R. B. Cattell (Ed.), *Handbook of multivariate experimental psychology.* Chicago: Rand McNally.

Cattell, R. B. (1988). Psychological theory and scientific method. In J. R. Nesselroade and R. B. Cattell (Eds.), *Handbook of multivariate experimental psychology* (2nd ed.). New York: Plenum.

Cattell, R. B., Cattell, A. K. S., and Rhymer, R. M. (1947). "P-technique demonstrated in determining psycho-physiological source traits in a normal individual," *Psychometrika.* **12,** 267–288.

Clarke, D. B., Clarke, A. M., and Brown, R. I. (1959). "Regression to the Mean—A Confused Concept," *British Journal of Psychology.* **51,** 106–117.

Clogg, C. C. (1981). "Latent structure models of mobility," *American Journal of Sociology.* **86,** 243–272.

Clogg, C. C. (1982). "Using association models in sociological research: Some examples," *American Journal of Sociology.* **88,** 114–134.

Coleman, J. S. (1964). *Introduction to mathematical sociology..* New York: Free Press.

Coleman, J. S. (1968). The mathematical study of change. In H. M. Blalock, Jr., and A. B. Blalock (Eds.), *Methodology in social research.* New York: McGraw-Hill.

Cronbach, L. J., and Furby, L. (1970). "How should we measure change—or should we?" *Psychological Bulletin.* **74,** 68–80.

Duncan, O. D. (1969). "Some linear models for two-wave, two-variable panel analysis," *Psychological Bulletin.* **72,** 177–182.

Etaugh, A. F., and Etaugh, C. F. (1973). "Change analysis is short-changed: A reply to Benjamin," *Developmental Psychology.* **8,** 184–186.

Featherman, D. L., and Petersen, T. (1986). "Markers of aging: Modeling the clocks that time us," *Research on Aging.* **8,** 339–365.

Furby, L. (1973). "Interpreting regression toward the mean in developmental research," *Developmental Psychology.* **8,** 172–179.

Garrett, H. E. (1946). "A developmental theory of intelligence," *American Psychologist.* **1,** 372–378.

Goldberger, A. S. (1964). *Econometric theory.* New York: Wiley.

Goldberger, A. S. (1971). "Econometrics and psychometrics: A survey of communalities." *Psychometrika.* **36,** 83–105.

Goodman, L. A. (1972). Some multiplicative models for the analysis of cross-classified data. In L. Le Cam et al. (Eds.), *Proceedings of the Sixth*

Berkeley Symposium on Mathematical Statistics and Probability, Vol. I. Berkeley: University of California Press.

Goodman, L. A. (1978). *Analyzing qualitative/categorical data, log-linear models, and latent structure analysis.* London: Addison-Wesley.

Hannan, M. T., and Tuma, N. B. (1979). Methods of temporal analysis. In A. Inkels, J. Coleman, and R. Turner (Eds.), *Annual review of sociology* (Vol. 5, pp. 303–328). Palo Alto: CA, Annual Reviews.

Hanushek, E. A., and Jackson, J. E. (1977). *Statistical methods for social scientists.* New York: Academic Press.

Harris, C. W. (1963). *Problems in measuring change.* Madison: University of Wisconsin Press.

Hauser, R. M., Dickinson, P. J., Travis, H. P. and Koffel, J. M. (1975a). "Structural changes in occupational mobility among men in the United States," *American Sociological Review,* **40,** 585–598.

Hauser, R. M., Dickinson, P. J., Travis, H. P., and Koffel, J. M. (1975b). "Temporal changes in occupational mobility: evidence for men in the United States," *American Sociological Review,* **40,** 279–297.

Heckman, J., and Singer, B. (1984). "Econometric duration analysis," *Journal of Econometrics.* **24,** 63–132.

Heise, D. R. (1970). Causal inference from panel data. In E. F. Borgatta and G. W. Bohrnstedt (Eds.), *Sociological methodology 1970* (pp. 3–27). San Francisco: Jossey-Bass.

Holtzman, W. H. (1963). Statistical models for the study of change in the single case. In C. W. Harris (Ed.), *Problems in measuring change.* Madison, WI: University of Wisconsin Press.

Hummel-Rossi, B., and Weinberg, S. L. (1975). *Practical guidelines in applying current theories to the measurement of change,* Parts I and II. Greenwich, CT: Johnson Associates.

Humphreys, L. G. (1961). Mimeographed paper on derived scores. Psychology Department, University of Illinois, Champaign-Urbana, IL.

Jones, C., and Nesselroade, J. R. (1989). *Multivariate, replicated, single-subject designs and P-technique factor analysis: A selective review of intraindividual change studies.* Unpublished manuscript.

Kenney, D. A. (1975). "Cross-lagged panel correlation: A test for spuriousness," *Psychological Bulletin.* **82,** 887–903.

Kessler, R. C. (1977). "The use of change scores as criteria in longitudinal survey research," *Quality and Quantity.* **11,** 43–66.

Kessler, R. C., and Greenberg, D. F. (1981). *Linear panel analysis: Models of quantitative change.* New York: Academic Press.

Labouvie, E. W. (1980). Identity versus equivalence of psychological measures and constructs. In L. W. Poon (Ed.), *Aging in the 1980s.* Washington, D.C.: American Psychological Association.

Labouvie, E. W. (1982). "Concepts of change and regression toward the mean," *Psychological Bulletin.* **92,** 251–257.

Lazarsfeld, P. F. (1940). "Panel studies," *Public Opinion Quarterly.* **4,** 122–128.
Lazarsfeld, P. F. (1948). The use of panels in social research. In P. F. Lazarsfeld, A. K. Pasanella, and M. Rosenberg (Eds.), *Continuities in the language of social research* (pp. 330–337). New York: Free Press. (Reprinted from *Proceedings of the American Philosophical Society*).
Lazarsfeld, P. F., Berelson, B. R., and Gaudet, H. (1944). *The people's choice.* New York: Duell, Sloan, and Pearce.
Lienert, G. A., and von Eye, A. (1987). "Nonparametric comparison of longitudinal response patterns from unpaired samples using configural frequency analysis," *Biometrical Journal.* **29,** 675–688.
Liker, J. K., Augustyniak, S., and Duncan, G. J. (1985). "Panel data and models of change: A comparison of first differences and conventional two-wave models," *Social Science Research.* **14,** 80–101.
Lord, F. M. (1963). Elementary models for measuring change. In C. W. Harris (Ed.), *Problems in measuring charge.* Madison, WI: University of Wisconsin Press.
Lytton, H., Croxen, M. E., and Pysh, F. (1973). "Regression to the mean misunderstood: A reply to Vockell and Asher," *Developmental Psychology.* **8,** 3–5.
Maxwell, S. E., and Howard, G. S. (1981). "Change scores—Necessarily anathema?" *Educational and Psychological Measurement.* **41,** 747–756.
McArdle, J. J. (1982). *Structural equation modeling of an individual system: Preliminary results from "a case study of episodic alcoholism."* Unpublished manuscript, Psychology Department, University of Denver, Denver, Colorado.
McArdle, J. J. (1988). Dynamic but structural equation modeling of repeated measures data. In J. R. Nesselroade and R. B. Cattell (Eds.), *Handbook of multivariate experimental psychology* (2nd ed.). New York: Plenum.
McNemar, Q. (1958). "On growth measurement," *Educational and Psychological Measurement.* **18,** 47–55.
Meredith, W., and Tisak, J. (1984). "Tuckerizing Curves." Paper presented at the annual meetings of the Psychometric Society, Santa Barbara, CA.
Molenaar, P. C. M. (1985). "A dynamic factor model for the analysis of multivariate time series," *Psychometrika.* **50,** 181–202.
Nesselroade, J. R. (1970). Application of multivariate strategies to problems of measuring and structuring long-term change. In L. R. Goulet and P. B. Baltes (Eds.), *Life-span developmental psychology: Research and theory.* New York: Academic Press.
Nesselroade, J. R. (1977). Issues in studying developmental change in adults from a multivariate perspective. In J. E. Birren and K. W. Schaie (Eds.), *Handbook of the psychology of aging.* New York: Van Nostrand Reinhold.
Nesselroade, J. R. (1983). Temporal selection and factor invariance in the study of development and change. In P. B. Baltes and O. G. Brim, Jr. (Eds.), *Life-span development and behavior* (Vol. 5). New York: Academic Press.

Nesselroade, J. R. (1986). "Selection and generalization in investigations of interrelationships among variables: Some commentary on aging research," *Educational Gerontology.* **12,** 395–402.

Nesselroade, J. R. (1988a). Sampling and generalizability: Adult development and aging research issues examined within the general methodological framework of selection. In K. W. Schaie, R. T. Campbell, W. Meredith, and S. C. Rawlings (Eds.), *Methodological issues in aging research.* New York: Springer.

Nesselroade, J. R. (1988b). Some implications of the trait-state distinction for the study of development over the life-span: The case of personality. In P. B. Baltes, D. L. Featherman, and R. M. Lerner (Eds.), *Life-span development and behavior* (Vol. 8). Hillsdale, NJ: Lawrence Erlbaum Associates.

Nesselroade, J. R., and Bartsch, T. W. (1977). Multivariate perspectives on the construct validity of the trait-state distinction. In R. B. Cattell and R. M. Dreger (Eds.), *Handbook of modern personality theory.*Washington, D.C.: Hemisphere.

Nesselroade, J. R., and Cable, D. G. (1974). " 'Sometimes, its okay to factor difference scores'—The separation of trait and state anxiety," *Multivariate Behavioral Research.* **9,** 273–282.

Nesselroade, J. R., and Ford, D. H. (1987). Methodological considerations in modeling living systems. In M. E. Ford and D. H. Ford (Eds.), *Human as self-constructing living systems.* Hillsdale, N.J.: Lawrence Erlbaum Associates.

Nesselroade, J. R., Pruchno, R. A., and Jacobs, A. (1986). "Reliability versus stability in the measurement of psychological states: An illustration with anxiety measures," *Psychologische Beitraege.* **28,** 255–264.

Nesselroade, J. R., Stigler, S. M., and Baltes, P. B. (1980). "Regression toward the mean and the study of change," *Psychological Bulletin.* **88,** 622–637.

Nunnally, J. C. (1967). *Psychometric theory.* New York: McGraw-Hill.

Orwin, R. G. (1981). "Measuring percentage change: Assumptions underlying unbiased treatment estimates," *Journal of Applied Psychology.* **66,** 671–676.

Overall, J. E., and Woodward, J. A. (1975). "Unreliability of difference scores: A paradox for measurement of change," *Psychological Bulletin.* **82,** 85–86.

Rao, C. R. (1958). "Some statistical methods for the comparison of growth curves," *Biometrics.* **14,** 1–17.

Rogosa, D. (1980). "A critique of cross-lagged correlation," *Psychological Bulletin.* **88,** 245–258.

Rogosa, D., Brandt, D., and Zimowski, M. (1982). "A growth curve approach to the measurement of change," *Psychological Bulletin.* **92,** 726–748.

Rogosa, D., and Willett, J. B. (1985). "Understanding correlates of change by modeling individual differences in growth," *Psychometrika.* **50,** 203–228.

Thorndike, R. L. (1942). "Regression fallacies in the matched groups experiment," *Psychometrika.* **7,** 85–102.

Tobin, J. (1958). "Estimation of relationships for limited dependent variables," *Econometrica.* **26,** 24–36.

Tornqvist, L., Vartia, P., and Vartia, Y. O. (1985). "How should relative change be measured?" *American Statistician.* **39,** 43–46.

Tucker, L. R. (1958). "Determination of parameters of a functional relation by factor analysis," *Psychometrika.* **23,** 19–23.

Tucker, L. R. (1966). Learning theory and multivariate experiment: Illustration by determination of parameters of generalized learning curves. In R. B. Cattell (Ed.), *Handbook of multivariate experimental psychology.* Chicago: Rand McNally.

Tucker, L. R., Damarin, F., and Messick, S. (1966). "A base-free measure of change," *Psychometrika.* **31,** 457–473.

Tuma, N. B., and Hannan, M. T. (1984). *Social dynamics: Models and methods.* New York: Academic Press.

Vockell, E. L., and Asher, W. (1973). "Methodological inaccuracies in Croxen and Lytton's reading disability and difficulties in finger localization and right-left discrimination," *Developmental Psychology.* **8,** 1–2.

von Eye, A., and Nesselroade, J. R. (1989, in press). "Types of change: Application of configural frequency to repeated observations in developmental research," *Experimental Aging Research.*

Wohlwill, J. F. (1973). *The study of behavioral development.* New York: Academic Press.

Chapter 2 Missing Data Estimation in Developmental Research

Michael J. Rovine and Mary Delaney*

College of Human Development
The Pennsylvania State University
University Park, Pennsylvania

Abstract

Methods of cross-sectional and longitudinal missing data estimation are presented. Since the selection of a technique depends on the pattern of missing data, different patterns are discussed. Both imputation methods (substitution of individual data points) and direct parameter estimation methods (estimation of variance/covariance matrices and mean vectors) are presented. Examples for two of the more general approaches, (1) cross-sectional maximum likelihood estimation of parameters using a multiple-group LISREL model, and (2) direct parameter estimation of longitudinal missing data in the presence of a nested pattern of missingness, are included.

1. Introduction

Possibly the most difficult estimation problem one has to deal with in developmental research is that of missing data. The statistical literature abounds with suggestions for handling missing data (Afifi and Elashoff, 1966; Little and Rubin, 1987), and at least one major statistical package (Dixon, 1983) has included procedures for estimating missing data, but

* The authors would like to thank Anne C. Petersen and Alexander von Eye for comments on an earlier draft.

few papers have appeared in the psychological and sociological literature (Kim and Curry, 1977). Papers dealing specifically with missing data in panel or longitudinal research (e.g., Marini, Olsen, and Rubin, 1980; Woolson and Clark, 1984) have also been limited. This, in part, may increase the tendency of researchers to make decisions regarding missing data estimation that are more dependent on what is generally used locally rather than on the basis of sound statistical principles. It is, therefore, the goal of this chapter to review some of the alternatives for missing data estimation.

An adequate solution to the missing data problem requires that a set of decisions be made. These are depicted schematically in Figure 2.1. The first decision depends on whether data points at the level of the item are missing on a scale or whether the whole scale is missing. If individual items are missing from a scale, present items may be used to estimate the missing items. These estimates can often be made independent of any considerations of statistical analyses. For example, if one item is missing on a 20-item scale, the scale can be contructed by substituting a value for that missing item and then analyzing the scale as if it were complete. The next decision concerns whether parameters in a statistical analysis can be estimated directly. As will be shown missing data can be handled either (1) by first estimating the missing points and then running the analysis as if complete data were present or (2) by directly estimating the parameter without estimating individual points. Because different methods may be needed depending on whether the data are cross-sectional or longitudinal in nature, a decision must be made whether to make cross-sectional or longitudinal estimates. To help make these decisions, we will discuss the following problems. First, estimation of individual items missing from a scale will be differentiated from problems that are specific to a particular analysis. Next, direct estimates of parameters without the imputation of individual missing data points will be discussed. Methods for first estimating missing values and then estimating parameters will then be discussed. Finally, problems met in panel or longitudinal studies will be addressed. Since the patterns of missing data determine which methods are most appropriate, these patterns will be established and described.

To illustrate the application of missing data principles, we have selected a set of variables from the Child–Family Development Project data set (Belsky, Gilstrap, and Rovine, 1984). Descriptive statistics for the selected variables are included in Table 2.1. Four subscales of the Braiker–Kelley marriage questionnaire (LOVE, CONFlict,

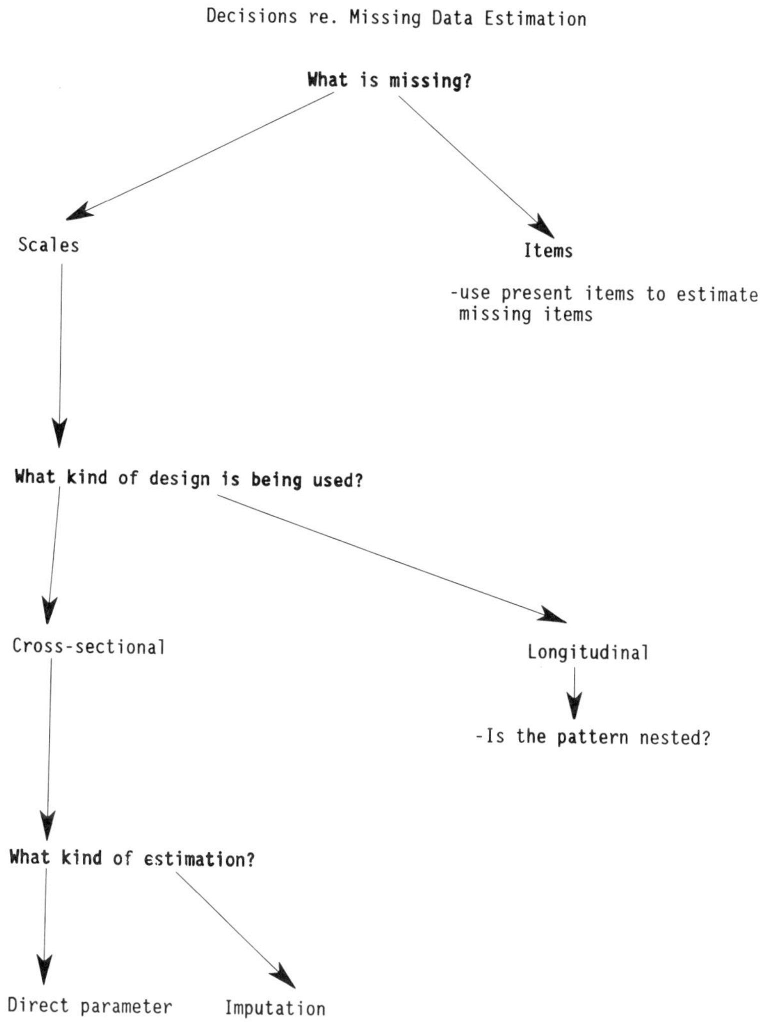

FIGURE 2.1. Decisions concerning missing data estimation.

AMBivalence, and MAINTenance), four subscales of the Jackson Personality Inventory (SELF-esteem, AFFect, NURTurance, and ENERGy), and four measures of temperament (FUSsy-DIStraught, UNADAPtive, DULL, and UNPREDictable) were selected. The marriage subscales were measured at three different waves of data collection (the prefix on the variable name refers to the wave).

TABLE 2.1. Descriptive Statistics for Wave 1 Subscales

Variable	N	Mean	Standard deviation
Love	158	78.04	9.14
Conflict	158	20.73	6.33
Ambivalence	158	10.56	5.52
Maintenance	158	29.96	6.18
Self-esteem	169	28.50	5.00
Affect	172	27.08	3.53
Nurturance	172	21.21	2.83
Energy	172	28.45	4.18
Unadaptive	170	8.50	3.29
Unpredictable	170	9.20	2.84
Dull	170	5.59	1.77
Fussy-Distraught	170	18.27	5.09

2. Estimating Scale Items

The first question to be considered is, under what circumstances can missing data be treated prior to any analysis? When one or a few items are missing from questionnaires or scales, these items can be estimated prior to scale construction and data analysis.

This kind of missing data often occurs because the respondent has overlooked or refused to answer an item. In this case the researcher need not delete all information for the individual on the scale. To properly adjust, the researcher must first determine how the scale was constructed. If the items are assumed to have been drawn from a single construct domain (Nunnally, 1978) to construct a k-item scale, the available items can be prorated to yield an estimated total scale value. If items are missing, the total X for the $k - a$ items can be adjusted by using the formula

$$X_{\text{adjusted}} = \frac{XK}{k - a}, \tag{1}$$

thus creating a total score using all available information but having the appropriate scale metric. This is equivalent to estimating the missing items by the mean of the individual's items on that scale.

If the questionnaire contains independent subscales, the adjustment should be based on only the items comprising the subscales from which an item is found missing. This process is equivalent to estimating the mean for the set of items reported and adding *a* items with scores at the mean to fill out the scale.

Inspection of the *N*'s for the variables suggests that three cases (possibly more) are missing items on one of the Jackson subscales. Inspection of the items shows that this is indeed the case. Under the assumption of single-domain scale construction, the items can be estimated from existent items on the same subscale.

When the items are missing because, for example, respondents refused to answer sensitive items, this procedure becomes questionable.

2.1. When Missing Data is a Problem

Unless sampling schemes are used that generate a predictable pattern of missing data, it is difficult to determine how severe the missing data problem might be. Certain circumstances exist in which the problem is trivial. Kim and Curry (1977) suggest that sample size, proportion of missing data, and the mechanism creating the missing data determine the degree to which it is not a problem. In addition, the type of analysis and number of variables used in a particular analysis determine the degree of problem.

If the proportion of missing data on any variable is low compared with the sample size, and if the data can be assumed to be missing completely at random (to be defined below), then any of the methods to be discussed below will yield results similar to those expected from a complete data analysis. However, a low rate of missingness may be tolerable in a univariate situation but may be problematic in a multivariate situation. Consider the situation involving 10 variables measured on 100 subjects. If 5% of the sample are missing values on a single variable, we can still have 95% cases with complete data. However, if 5% of the sample have missing data on each variable, and the pattern of missing data is random (i.e., one cannot predict who is missing a value on variable 10 by knowing who is missing the first nine variables) and there is little overlap of missing values, then as many as half the cases may be candidates for deletion in any analysis that requires all 10 variables.

There is always, of course, a reason why a data point is missing. If this reason can be linked to any measured variable or characteristic; that is, if there is something systematic about the missing point, that characteristic should be used to make an estimate of its data point. If not, then the group parameter estimates can be used to estimate the missing point. Regression methods for these estimations are discussed below.

3. Patterns of Missing Data

The first step in any analysis of missing data should involve the determination of the pattern of missing data answering the question of which individuals are missing values on which scales. However, the pattern can also provide valuable information needed to select an estimation method. The basic strategy for determining patterns can be understood if one thinks of each variable in a data set as having a companion dichotomous variable representing whether the variable is missing or not. These new variables can be used to check whether associations (e.g., correlations) exist between the pattern of missing (1 = present, 0 = missing) and other variables. Different kinds of associations have been described by Rubin (1976).

For a single variable X, data are said to be *missing at random* (MAR) if the pattern of missing data is not dependent on values of X. If X is a measure of education, X would not be "missing at random" if those with a low degree of education refuse to report their education.

For a set of variables $X_1, \ldots, X_k$, each of which has missing values, data are missing at random if the pattern of missing data is independent of any of the members of that set.

For a single variable X, data are said to be *observed at random* (OAR) if the pattern of missing data is independent of values of *other* observed variables. If, then, in our example above, there is no relation between amount of education and whether education is missing, education would be missing at random. If there were an association between whether education was missing and another observed variable, SAT score, then the data would not be observed at random.

Rubin (1976) defines variables that are both MAR and OAR as "missing completely at random" (MCAR). Whether such variables are missing is independent of any information in the data set. As such, those with missing values on an MCAR variable are assumed to comprise a random subsample of the original sample. In this case the mechanism that is causing the missing data pattern can be ignored (Allison, 1982).

When the missing data structure is caused by an underlying but unmeasured variable, a proxy measure that is related to the underlying cause of the missing data can be used to improve the missing data estimate. If, for example, nonresponse depends on OCCUPATION, but OCCUPATION has not been measured, one might use another variable, INCOME, that is closely related to improve the estimate.

The more difficult problem occurs when MAR is violated and the incidence of missing data depends on the value of the variable that is missing. One example would have individuals with personality disorders refuse to complete a personality inventory. In this situation complex models for nonignorable missing data are available (Little and Rubin, 1987).

3.1. Establishing the Pattern of Missing Data

Since the estimation techniques to be discussed assume MAR data, tests of the assumption would seem necessary. However, an exact test would require the missing values to determine if indeed there was an association between the values of a variable and whether or not it was missing. If this information were available, then, of course, the data would not be missing. Complex sampling techniques (e.g., double sampling) have been suggested to give the investigator some control over the structure of missing data.

Tests have been suggested by Kim and Curry (1977) and Cohen and Cohen (1975). Both use a dichotomous missing data indicator for each variable in the data set. Log-linear models can be used to test whether the incidence of missing data on one variable predicts the incidence of missing data on other variables (Kim and Curry, 1977). This is not, however, a test of the MAR pattern, but only of whether an association exists between different patterns of missing data. Some indication of the MAR pattern is suggested by the independence of the missing data structures, but, as Allison (1982) points out, certain patterns that can be easily dealt with (e.g., patterns of variables missing in sets) would be treated as not MAR when, in fact, they can be handled by maximum likelihood techniques.

One way of getting some indication of the degree to which a pattern is MAR is to test the more restricted pattern of MCAR. If the pattern of missing data is MCAR, then each subset of how the sample represented by a different pattern of missing data should represent a random subsample. If this is the case, no differences would be expected on other variables between missing data subsamples and the complete data subsample on other variables (e.g., demographics) measured. Any differences that might occur would point in the direction of a possible mechanism for the existence of missing data. Since MCAR is a necessary, but not a sufficient condition for MAR, such a test could at best indicate that MAR might be a reasonable assumption.

Practically speaking, MAR is more often assumed than tested.

Methods are available to deal with missing data when the pattern is not MAR (Little and Rubin, 1987), but these methods are normally very difficult to compute and usually restricted to that particular analysis.

Indicator variables can handle questions about the assumption of OAR. This information can be helpful in selecting which variables to use in many of the regression imputation procedures to be discussed. OAR requires that the pattern of missing data on one or a set of variables be independent of observed values on other variables. The indicator variables (1 = present, 0 = missing) can be added as an additional effect or predictor in analyses to determine whether the relationship expressed by an analysis depends on the pattern of missing data. For example, indicator variables for one variable can be used to look at mean differences in other variables.

In large studies with spotty missing data on a number of variables, many different patterns of missing data may exist. To determine whether these can be reduced to a few prototypical patterns, Cohen and Cohen (1975) suggest searching for common patterns among the indicator variables using factor-analytic techniques. Factors would represent clusters of variables that tended to be missing on the same individuals.

A correlation matrix of the dichotomous indicator variables is shown in Table 2.2. As can be seen patterns of missingness are essentially the same within subscales and uncorrelated across subscales.

TABLE 2.2. Correlations of Dichotomous Indicator Variable

Marriage questionnaire												
W3LOVE	1.00											
W3CONF	1.00	1.00										
W3AMB	1.00	1.00	1.00									
W3MAINT	1.00	1.00	1.00	1.00								
Personality inventory												
PWJSELF	−.05	−.05	−.05	−.05	1.00							
PWJAFF	−.02	−.02	−.02	−.02	1.00	1.00						
PWJNURT	−.02	−.02	−.02	−.02	1.00	1.00	1.00					
PWJENRG	−.02	−.02	−.02	−.02	1.00	1.00	1.00	1.00				
Temperament												
W3UNADAP	−.04	−.04	−.04	−.04	−.02	−.01	−.01	−.01	1.00			
W3UNPRED	−.04	−.04	−.04	−.04	−.02	−.01	−.01	−.01	1.00	1.00		
W3DULL	−.04	−.04	−.04	−.04	−.02	−.01	−.01	−.01	1.00	1.00	1.00	
W3FUSDIS	−.04	−.04	−.04	−.04	−.02	−.01	−.01	−.01	1.00	1.00	1.00	1.00

In longitudinal or panel designs, the optimal pattern of MCAR cannot be assumed because, in addition to whatever spotty missing data might occur, whole waves may be missing. In this case of missing waves of data, nested patterns (Marini, Olsen, and Rubin, 1980) have been shown to lead to adequate parameter estimates. These patterns will be discussed in more detail below.

In addition to analyzing dichotomous missing data patterns of single variables, one can construct a variable that combines missing data indicators of multiple variables. For example, imagine three variables $x, y,$ and z. If they are recoded as dichotomous indicator variables (M_1, M_2, and M_3), eight combinations can result. A single missing pattern variable can be created by

$$\text{MISINDEX} = 100M_1 + 10M_2 + M_3. \tag{2}$$

This index can be used to generate a frequency distribution showing which patterns of missing data are most prevalent. In a single wave of data, this index can quickly show how many cases are lost in a complete case analysis with different combinations of variables. It can also indicate how many missing data points must be estimated in an analysis involving a certain subset of variables. In a multiwave design, this index can be used to determine how many cases can be used in any repeated measures analysis requiring complete data across waves. In a multiwave design, tests of the pattern of missingness can be achieved by using methods that look for associations among the missing indicator variables (e.g., log-linear analysis). Tests of hypotheses concerning these patterns using configural frequency analysis (von Eye, Chapter 16, this volume) are also available.

3.2. *The Problem of Simultaneously Estimating Missing Data for Correlations and Group Differences*

Once an optimal method (for criteria, see below) has been chosen to estimate missing data, that method could be used to fill in the blanks for each missing value in the raw data set. This, however, can get the researcher into serious trouble, because what may be a good missing data estimate for one analysis may not work well for another. An estimate that is unbiased for the first moment of a distribution may be biased for the second moment.

To illustrate one possible problem we present the following data:

Subject	X_1	X_2
1	1	3
2	3	5
3	5	7
4	4	—
5	7	9

Two variables, X and Y, are measured on five individuals. Person 4, however, is missing the value for variable X. If the goal is to estimate the missing data point and maintain the parameter estimates, the question becomes which parameter estimate? To maintain the mean for X, one should choose the value $X_4 = 4$. However, that value would change the estimate of the correlation from $r = 1.00$ to $r = 0.71$. To maintain the correlation, choose the value $X_4 = -1$. This value, however, changes the mean of X to $X = 3$. One of the two parameters will be biased.

3.3. Estimating Parameters Directly

When analyzing data in which some of the variables have some missing values, one can often use a summary form of the data (i.e., a matrix of covariances and means) to estimate parameters of interest. This is especially true since many of the estimates can be based on the general linear model. However, the adequacy of these summary statistics depends, in part, on the occurrence and pattern of missing data. If any data missing can be assumed to be MCAR, the summary statistics can be considered unbiased estimates (Little and Rubin, 1987) of the population parameters, and other parameters (e.g., regression weights) can be estimated on the basis of those summary statistics. In addition, data are often reanalyzed with different subsets of variables. In this case, the process of generating summary statistics, estimating parameters, and then testing for the adequacy of the parameters can be cumbersome. What may then be an adequate set of summary statistics for one set or subset of variables in a particular analysis may prove inadequate for another subset in a subsequent analysis. Thus, for archival purposes and for ease of analysis, it often seems reasonable to insert individual missing data points and then treat the analysis as if complete data were there. This means that the decision must be made whether to estimate parameters directly or to first insert actual missing data values and then estimate the parameters of interest.

If the amount of missing data is small relative to the amount of data collected, and if the pattern of missing data is MCAR, the parameters of interest may be estimated directly under the assumption that cases deleted would not substantially change the estimates. We will begin the discussion of missing data estimation techniques by presenting what have traditionally been the most common methods of direct parameter estimation: listwise and pairwise deletion (Nie et al., 1975).

If there is a fair amount of missing data and a number of different analyses to be performed, a two-step procedure may be necessary: missing data points are first estimated; then the parameters are estimated. After discussing the more common methods, mean and subgroup mean insertion, variance-maintaining estimation, hot-deck procedures, regression procedures, and expectation maximization (EM) algorithms will be considered.

3.4. Cross-Sectional versus Longitudinal Data

Some additional considerations are necessary when the data set is longitudinal in nature.

1. The pattern of missing data must reflect both spotty missing data at each time of measurement and systematic problems such as dropouts.
2. The variables that can act as indicators of the cause of missing data must be appropriately selected from prior occasions and from the same occasion. We will see how certain patterns of longitudinal data can be easily handled using likelihood techniques.

3.5. Properties of Missing Data Estimates

As with all estimation, the adjustment for missing data should yield unbiased estimates of population parameters. In the case of MCAR data, the estimate is unbiased, because those with missing data represent a random subsample of the sample considered. If, however, there is a systematic mechanism causing missing data that would tend to bias the estimate up or down, the missing data estimate should adjust the parameter estimate to remove nonresponse bias (Little and Rubin, 1987). The degree to which this is accomplished depends on how well that mechanism is known.

The variance of the sample distribution ideally should also be maintained. Any decrease in the variance will affect the relationship (correlation or covariance) between that variable and any other.

As with all parameter estimates, optimal missing data estimates should be efficient and consistent. Efficient estimates have the property that the standard error of the estimate is as small as possible. When these are tested for statistical significance, efficient estimates are most likely to appear significant. Consistent estimators have the characteristic that as the sample size increases the expected value of the estimate gets closer to the value of the parameter. Thus consistent estimates are asymptotically unbiased.

4. Cross-Sectional Methods

Techniques for analyzing cross-sectional data sets with missing data along with their major benefits and drawbacks are presented in Table 2.3. If the amount of missing data is small and can be assumed to be missing completely at random, parameters can be directly estimated without the prior estimation of individual points. Two methods have been suggested: listwise and pairwise deletion of cases with missing data.

4.1. Listwise Deletion versus Pairwise Deletion

Listwise deletion requires that all cases with variables missing be dropped from the analysis. Although listwise deletion can remove large amounts of data from an analysis, the advantages reflect the well-behaved nature of the listwise estimates. When data are MCAR, the reduced data set represents a randomly drawn subsample of the original data set. Inferences about the values of the population parameters can thus be considered reasonable. Pairwise deletion, on the other hand, calculates a correlation or covariance based on all cases with data on each pair of variables. Although the sample size can vary with each calculation, pairwise deletion makes use of more of the available data.

The choice between these two methods has been a matter of controversy. Haitovsky (1968) concluded that listwise deletion was superior to pairwise deletion in the estimation of partial regression coefficients when the pattern of missing data was random and the proportion of data missing was relatively small. Kim and Curry (1977) showed, however, that for moderate correlations (less than .70) pairwise

TABLE 2.3. Methods of Missing Data Estimation for Cross-Sectional Studies

Method	Assets	Liabilities	Pattern
Direct parameter estimation			
Listwise deletion	Gives relatively unbiased estimates if number of missing values is small or if variables are highly correlated. Generates consistent matrices	Removes too many cases Gives biased estimates when pattern is nonrandom	Missing completely at random (MCAR)
Pairwise deletion	Makes use of more of the data	Can generate inconsistent correlation and covariance matrices	MCAR
Multiple-group maximum likelihood estimation	Estimates are consistent, efficient, and normally distributed	Requires ML estimation procedure	MCAR within each subgroup
Mean insertion	Can generate unbiased estimates of means for ANOVA	Shrinks variance Affects correlations Generates small error term for ANOVA	Completely random missing data
Subgroup mean insertion	Maintains some individual difference variation	Can generate biased estimates Dependent on the selection of groups	Random missing data
Mean with stochastic term	Maintains mean and variance structure	Can generate some strange data points Consistent matrices can have small eigenvalues	Completely random missing data
Subgroup mean with stochastic term	Combines 4 and 5		Random missing data
Regression estimates	Sensitive to individual differences	Dependent on variable selection	Random missing data
Expectation maximization (EM) iteration methods	Maximizes number of individual difference variables used in estimation	Hard to compute May function no better than simpler methods	Random missing data

deletion could give estimates closer to the original correlation matrix when the proportion of missing data was as high as 10% and as low as 1%.

The characteristics of the estimates also lead to an uncertain conclusion. Pairwise deletion produces estimates that are consistent (Glasser, 1964). However, Allison (1982) points out that these estimates may be less efficient than listwise deletion. Higher correlations seem to yield higher relative efficiency when listwise deletion is used. Although the estimates using pairwise correlations are consistent, the estimates of the standard errors of those correlations are not (Allison, 1982). Since significance tests using uncorrected standard errors may yield misleading results, formulas for the adjustment of these tests are available (Glasser, 1964).

Often the critical choice in deciding between listwise and pairwise deletion concerns characteristics of the covariance matrices generated by both methods. The covariance matrix generated by listwise deletion will always be consistent (i.e., positive semidefinite; Anderson and Gerbing, 1984). If, however, there is a systematic mechanism causing specific variable values to be missing, the estimates of the parameters will be biased. A major advantage, then, of listwise deletion is that any analysis that depends on the inversion of a matrix can be computed.

When different variables have different numbers of missing data points, pairwise deletion will make use of more available data. A covariance matrix can, for example, have n's that vary with each pair of variables. When only a few of the variables in the data set have many missing data points, the n with listwise deletion will not reflect the amount of data available. Pairwise deletion computes a covariance matrix in which each covariance is based on all cases on which each pair of variables has data. If the pattern of missing data is not random, an inconsistent matrix (not positive-definite) can result. This matrix can only be inverted using a generalized inverse and thus cannot be used in many multivariate analyses.

An inconsistent matrix is most likely to occur either as the n for the two variables comes closer to the number of variables in the matrix or if some subset of the matrix has correlated missing data indicators. An attempt at matrix inversion can thus serve as a quick test of whether the pattern of missingness should be of immediate concern. If the matrix cannot be inverted, a generalized inverse can be computed on the pairwise matrix (Rao and Mitra, 1971).

If the subgroups with specific missing data patterns are large enough,

TABLE 2.4. Pairwise and Listwise Correlation Matrices

	Pairwise correlation matrix											
Love	1.00											
Conflict	−.47	1.00										
Ambivalence	−.72	.56	1.00									
Maintenance	.45	.04	−.19	1.00								
Self-esteem	−.05	−.03	.00	−.16	1.00							
Affect	−.14	.04	.10	−.17	.03	100						
Nurturance	−.14	−.01	.10	−.20	.01	.54	1.00					
Energy	−.07	.10	.09	−.02	.48	.02	.07	1.00				
Unadaptive	−.12	.13	.16	−.01	.10	.09	.08	.24	1.00			
Unpredictable	−.12	−.04	.08	−.17	.06	.09	.09	.09	.11	1.00		
Dull	−.05	.10	.06	.10	.02	−.06	−.12	.05	.01	.01	1.00	
Fussy-distraught	−.11	.14	.06	−.09	.00	−.04	.07	.14	.33	.30	.32	1.00
	Listwise correlation matrix ($N = 148$)											
Love	1.00											
Conflict	−.44	1.00										
Ambivalence	−.72	.54	1.00									
Maintenance	.50	.03	−.21	1.00								
Self-esteem	−.04	−.08	.00	−.16	1.00							
Affect	−.14	.05	.12	−.19	.06	1.00						
Nurturance	−.14	.00	.11	−.24	.01	.55	1.00					
Energy	−.07	.10	.10	−.04	.50	−.02	.06	1.00				
Unadaptive	−.09	.10	.12	−.04	.11	.11	.13	.25	1.00			
Unpredictable	−.13	−.01	.08	−.13	.11	.08	.16	.12	.10	1.00		
Dull	−.08	.07	.09	.11	.00	−.09	−.14	.03	.04	.04	1.00	
Fussy-distraught	−.11	.13	.02	−.08	.02	−.07	.11	.14	.32	.29	.35	1.00

multiple-group models can be used to determine which subgroups can be combined for pairwise deletion. If, however, the different subgroups cannot be combined, then either listwise deletion or the individual analysis of the subgroups become the options for direct parameter estimation.

Pairwise and listwise correlation matrices for our sample data set are included in Table 2.4. As can be seen, the two sets of estimates are very similar. There are, however, some discrepancies, so choosing a matrix does represent a somewhat different modeling of the data.

4.2. Multiple-Group Maximum Likelihood Estimation

Since problems occur with both the listwise or pairwise methods, Allison (1982) suggested an alternative using multiple-group maximum likelihood (ML) estimation to determine whether patterns of covariation are similar for subgroups with different missing data structures. These estimates essentially are selected to maximize the probability that the resulting estimates will reflect what has already been observed in the sample. The ML estimates have been shown to be consistent, efficient, and normally distributed. The methods to be described will work best when subgroups with the same missing data structures are large enough to provide stable estimates of zero-order correlations or covariances.

Unlike pairwise and listwise methods, which yield a correlation or covariance matrix, Allison's method goes directly to the estimation of general linear model parameters. We now illustrate how this method would work in the estimation of regression parameters.

Allison (1982) proposed a direct parameter estimation method using ML estimation based on LISREL's (Joreskog and Sorbom, 1984) capacity to estimate multiple-group models in which constraints on the solution across groups can be specified. Such a model can estimate a set of parameters that are common to all missing data structures (each represented by a different group) while testing how these common estimates account for the patterns of covariation in each of the subgroups.

Consider the situation in Figure 2.2, in which a path model is used to represent a set of regression relationships among our example variables.

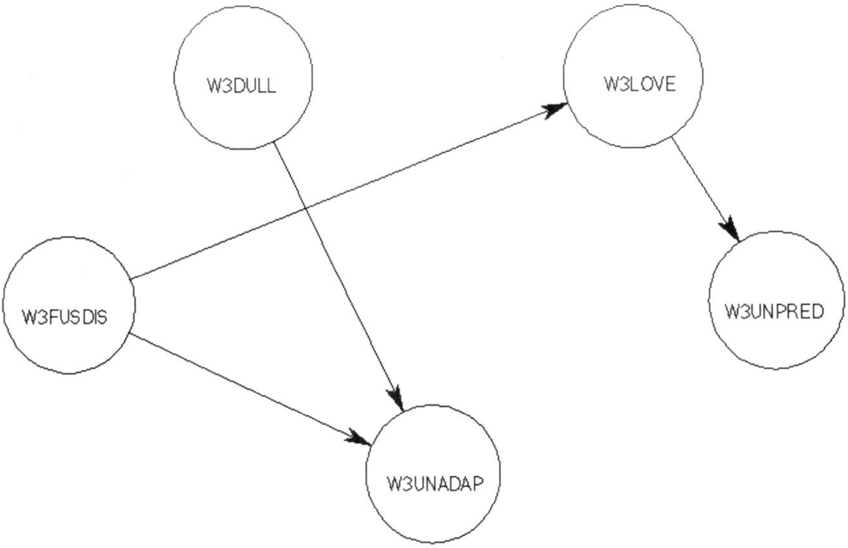

FIGURE 2.2. Regressions relationships estimated by multiple—group LISREL model.

The relationships described by the path diagram can be analyzed with the following regression equations:

$$Y_{3i} = B_{31} Y_{1i} + e_{3i}, \tag{3}$$

$$Y_{4i} = B_{42} Y_{2i} + B_{41} Y_{1i} + e_{4i}, \tag{4}$$

$$Y_{5i} = B_{53} Y_{3i} + e_{5i}. \tag{5}$$

If a correlation or covariance matrix has been computed for those six variables, the relationships necessary to estimate the model are shown schematically as

	Y_1	Y_2	Y_3	Y_4	Y_5
Y_1					
Y_2	X				
Y_3	X				
Y_4	X	X			
Y_5			X		

Other off-diagonal correlations are not essential for estimating the model. If the missing data subgroups are functioning under the same model, the same structure should hold for each subgroup.

Now assume that a large subgroup is missing variable Y_3. That group's matrix would be

	Y_1	Y_2	Y_3	Y_4	Y_5
Y_1					
Y_2	X				
Y_3	0				
Y_4	X	X			
Y_5			X		

Note that the correlation between Y_3 and Y_2 is set to zero. To test the assumption that the same path model holds for both subgroups, one can use a two-group LISREL model. To do this, we must rewrite the regression equations as latent variable equations. By specifying a set of latent variables η_1, η_2, η_3, η_4, η_5, each having a single indicator, this can be done. The model for a single group becomes

$$\boldsymbol{\eta} + \mathbf{B}\boldsymbol{\eta} + \boldsymbol{\zeta}, \tag{6}$$

where $\boldsymbol{\eta}$ is the matrix of latent variables, $\mathbf{B}$ is the matrix of regression weights, and $\boldsymbol{\zeta}$ is the matrix of residuals.

The indicators can be linked to the latent variables by

$$\mathbf{Y} = \boldsymbol{\Lambda}_Y \boldsymbol{\eta} + \boldsymbol{\varepsilon}, \tag{7}$$

where $\mathbf{Y}$ is the matrix of observable variables, $\mathbf{\Lambda}_Y$ is an identity matrix, and $\boldsymbol{\varepsilon}$ is a matrix of residuals.

To compute estimates, Allison (1982) suggests constraining the model parameters equally across groups, with the exception that, in the missing data group, $\lambda_{33} = 0$ and $\text{var}(e_e) = 1.0$, representing the missing variable Y_3 in that group.

Covariance matrices are generated for both groups, with a complete matrix for the complete data group and a matrix with 0.0 representing the correlation or covariances of all variables with the missing variable Y_3 in the missing data group. If a covariance matrix is used, the variance of Y_3 is scaled to 1.0 in the missing data group.

For our data example we selected five of the first-wave variables: W3FUSDIS, W3DULL, W3LOVE, W3UNADAP, W3UNPRED. Four of the variables have complete data. Fifteen individuals are missing data on W3LOVE. To estimate regression weights for this data, we need different pattern matrix specifications for each group. These specifications are

Group 1 (complete data)

$$\mathbf{\Lambda}_Y = \begin{bmatrix} 1 & 0 & 0 & 0 & 0 \\ 0 & 1 & 0 & 0 & 0 \\ 0 & 0 & 1 & 0 & 0 \\ 0 & 0 & 0 & 1 & 0 \\ 0 & 0 & 0 & 0 & 1 \end{bmatrix}, \qquad \boldsymbol{\theta}_\varepsilon = [\ 0 \quad 0 \quad 0 \quad 0 \quad 0\ ]. \tag{8}$$

Group 2 (missing data)

$$\mathbf{\Lambda}_Y = \begin{bmatrix} 1 & 0 & 0 & 0 & 0 \\ 0 & 1 & 0 & 0 & 0 \\ 0 & 0 & 0 & 0 & 0 \\ 0 & 0 & 0 & 1 & 0 \\ 0 & 0 & 0 & 0 & 1 \end{bmatrix}, \qquad \boldsymbol{\theta}_\varepsilon = [\ 0 \quad 0 \quad 1 \quad 0 \quad 0\ ].$$

Both groups

$$\mathbf{B} = \begin{bmatrix} 0 & 0 & 0 & 0 & 0 \\ 0 & 0 & 0 & 0 & 0 \\ 1 & 0 & 0 & 0 & 0 \\ 1 & 1 & 0 & 0 & 0 \\ 0 & 0 & 1 & 0 & 0 \end{bmatrix}, \qquad \boldsymbol{\psi} = \begin{bmatrix} 0 & & & & \\ 0 & 0 & & & \\ 0 & 0 & 1 & & \\ 0 & 0 & 0 & 1 & \\ 0 & 0 & 0 & 0 & 1 \end{bmatrix}.$$

By specifying the estimates for all matrices as invariant (i.e., BE = IN, PS = IN, etc.), we can generate a common set of regression estimates and the accompanying test of that solution for both groups. The results of this analysis appear in Table 2.5. A common set of regression estimates has thus been generated for both groups. Inspection of the input correlation matrices for both groups shows large differences in the sizes of some of the correlations. This is due partly to group differences and, partly to the small sample size for the missing data group. Although the ML method is expected to weight estimates for group size, the adequacy of the weighting can be observed by comparing the results obtained against the results one would get by analyzing only the data for the complete data group. One way to accomplish this would involve running the model again and asking for separate estimates for the subgroups. Using the

TABLE 2.5.

Group 1: *complete data group*

$$\lambda_y = \begin{bmatrix} 1 & 0 & 0 & 0 & 0 \\ 0 & 1 & 0 & 0 & 0 \\ 0 & 0 & 1 & 0 & 0 \\ 0 & 0 & 0 & 1 & 0 \\ 0 & 0 & 0 & 0 & 1 \end{bmatrix} \qquad \beta = \begin{bmatrix} 0 & 0 & 0 & 0 & 0 \\ 0 & 0 & 0 & 0 & 0 \\ -.12 & 0 & 0 & 0 & 0 \\ .36 & -.11 & 0 & 0 & 0 \\ 0 & 0 & -.14 & 0 & 0 \end{bmatrix}$$

$$\psi = \begin{bmatrix} 1 & & & & \\ 0 & 1 & & & \\ 0 & 0 & .99 & & \\ 0 & 0 & 0 & .88 & \\ 0 & 0 & 0 & 0 & .98 \end{bmatrix} \qquad \theta_\varepsilon = \begin{bmatrix} 0 & & & & \\ 0 & 0 & & & \\ 0 & 0 & 0 & & \\ 0 & 0 & 0 & 0 & \\ 0 & 0 & 0 & 0 & 0 \end{bmatrix}$$

Goodness-of-fit index: .93

$$\lambda_y = \begin{bmatrix} 1 & 0 & 0 & 0 & 0 \\ 0 & 1 & 0 & 0 & 0 \\ 0 & 0 & 1 & 0 & 0 \\ 0 & 0 & 0 & 1 & 0 \\ 0 & 0 & 0 & 0 & 1 \end{bmatrix} \qquad \beta = \begin{bmatrix} 0 & 0 & 0 & 0 & 0 \\ 0 & 0 & 0 & 0 & 0 \\ -.12 & 0 & 0 & 0 & 0 \\ .36 & -.11 & 0 & 0 & 0 \\ 0 & 0 & -.14 & 0 & 0 \end{bmatrix}$$

$$\psi = \begin{bmatrix} 1 & & & & \\ 0 & 1 & & & \\ 0 & 0 & .99 & & \\ 0 & 0 & 0 & .88 & \\ 0 & 0 & 0 & 0 & .98 \end{bmatrix} \qquad \theta_\varepsilon = \begin{bmatrix} 0 & & & & \\ 0 & 0 & & & \\ 0 & 0 & 1 & & \\ 0 & 0 & 0 & 0 & \\ 0 & 0 & 0 & 0 & 0 \end{bmatrix}$$

Goodness-of-fit index: .81

$\chi^2(23) = 42.26$ ($p = .008$)

pattern matrices listed above, one can request separate estimates for the **B** and ψ matrices (i.e., LY = IN, TE = IN, BE = PS, PS = PS). Complete LISREL specifications for the model are included in the appendix.

4.3. Estimation of Missing Data Points

If the number of variables in a data set is large and the researcher wishes to run a set of analyses on a common data set, estimation of missing data points is often considered. Methods to be considered below first estimate missing data points and then proceed to conduct the analyses as if the data were complete.

4.3.1. Means. Perhaps the most common way to estimate a missing data point for a variable involves using the mean of that variable computed for all observations with a value for that variable. Since for

$$\bar{x} = \sum_{i=1}^{N} \frac{x_i}{N} \tag{9}$$

the expected value

$$E(x) = \mu, \tag{10}$$

the estimate is unbiased. The variance of X_i, however, will shrink, since all values of X_i added will contribute deviations of $(X_i - X)^2 = 0$ to the variance. Estimating the missing value by the mean serves to increase the denominator of the variance while keeping the numerator the same.

The use of such an estimate will affect the correlation between that variable and any other by decreasing its variability. In ANOVA, use of such estimation will serve to decrease the variability within each cell and shrink the pooled error term. Error degrees of freedom can be adjusted to somewhat improve the statistical test. (One degree of freedom can be subtracted from the error for each inserted data point.) In addition, frequency distributions with the now complete data set may be misleading because too many values are now located at the center, creating a more leptokurtic distribution.

One way to maintain some of the variance that real variable values would be expected to have involves using subgroup means as estimates. If a categorical variable can be found that is related not to whether the data point is missing but to the value of the missing datum itself, the sample can be subdivided according to that variable and a means can be computed for each subgroup.

4.3.2. Means with a Stochastic Term. To maintain the variance of the complete case subsample, we can add a stochastic term to either the mean or the subgroup mean estimate. The new estimate would take the form

$$X_{i_{\text{miss}}} = \bar{x} + cr, \tag{11}$$

where c represents a constant that can be chosen to maintain the variance, and r is a random number. When the data are MCAR, these will yield estimates of means as adequate as the simpler mean imputation. The estimates of the variance will be closer to the estimated population variance; however, covariances will still be underestimated. When a systematic reason for the data to be missing exists, these estimates will be biased.

4.3.3 Buck's Methods. Buck (1960) suggested separate regressions for each different pattern of missing data to generate regression estimates to compute a missing data value. For each missing variable, those cases with a common pattern of missingness and a common set of predictor variables are determined. A missing value set of cases with all predictors but a missing value for the selected variable is also identified. The variable with missing data is then regressed onto the predictor variables in the complete cases set. The regression weights generated are then used to estimate the missing values in the other data set. This method requires that a different regression equation be used for each pattern of missing data. Results from this method are reported to yield good estimates of the means, to underestimate the variances and covariances somewhat less than the methods mentioned above, and to yield frequency distributions that are somewhat less leptokurtic (Little and Rubin, 1987).

4.3.4. Other Regression Methods. Buck's method requires using every available variables as a predictor; other regression methods try to select a set of predictors that may better represent the underlying cause of the missing data structure. These methods select those variables that best predict the missing variable and compute regression coefficients that are used to estimate the missing data value. These methods typically will yield good estimates of the means, and will tend to underestimate variances and covariances. To the extent that the mechanism underlying the missing data structure is reflected in the predictor variables, these methods can be expected to yield less biased results.

4.3.5. Hot-Deck Methods. Hot-deck methods (Little and Rubin, 1987) assume that the distribution, and not just a parameter estimate derived from the distribution, is the best source for a missing data estimate. To implement this strategy, one first generates a frequency distribution based on those cases with values for variable. Each possible variable value is then weighted by its frequency. Missing values are replaced by selecting at random from a "deck" of "cards." The number of cards in the deck equals the cumulative frequency, and the "face value" of each card equals the variable value. This is equivalent to selecting values from a hypergeometric distribution.

This method will tend to give somewhat biased estimates of the means. Variance estimates, which tend to be larger than the complete case population variance, can be adjusted (Little and Rubin, 1987).

4.3.6. Principal Components Methods. With many variables and a small sample, it has been suggested that principal components can be computed on a listwise deletion correlation matrix. The weights generated can be used to recover the raw data. The transformations involved can also be used to estimate missing data points. Simulations (Timm, 1970) have suggested that this method is generally inferior to regression methods except when the number of variables is large in comparison to the sample size and there is a large amount of missing data. However, with a large amount of missing data, listwise deletion as a required first step is questionable.

4.3.7. Logistic Regression Models. Greenlees, Reece, and Zieschang (1982) suggest a logistic regression model be used to estimate missing data points when the existence of the missing point is related to the value of the missing variable. Their method first orders the data into respondents and nonrespondents and then factors the likelihood function. The contribution of each respondent to the likelihood is the product of the probability of the response given the variable and the value of the probability density function for that variable for the particular value. For each of the nonrespondents the contribution is the marginal probability of nonresponse. The joint product of these contributions can then be used to generate the estimates.

4.4. Likehood-Based Estimation (EM Algorithms)

Of the likelihood-based procedures available, probably the most general and most easily available (BMDP, 1985) are the EM algorithms (Orchard and Woodbury, 1972). These methods are iterative in the sense that they use cases with complete data only to estimate missing values, then those values are plugged in and used to reestimate the parameters. Missing values are then reestimated on the complete-data-only subset by using the new parameter estimates. The process is repeated until the parameters converge. The problem with these estimates has traditionally been that because they use ML estimation in the maximizing step, they are difficult to obtain without a computer. However, major packages routinely include at least some ML procedures (Dixon, 1983; SAS Statistics Manual, 1985).

The likelihood function $L(\theta)$, gives the probability that the values of a variable X observed in a sample would give the parameter θ. Given a set of n observed values, $X = (x_1, x_2, \ldots, x_n)'$, where x_1^t represents the current value of the estimate for x_1 the likelihood that these values would be observed given θ is

$$L(\theta) = \prod_{i=1}^{n} f(x_i \mid \theta). \tag{12}$$

For computational reasons, the log of the likelihood is often used and takes the form

$$l(\theta) = \sum_{i=1}^{n} \ln(f(x_i \mid \theta)). \tag{13}$$

The E step calculates the expected values of the missing data given the parameter estimates and observed data. The M step then estimates θ by maximizing the log-likelihood for θ using the estimated missing data (Little and Rubin, 1987).

If, for example, X was normally distributed with $\theta = (\mu, \sigma^2)$ and m nonmissing values, the E step would calculate

$$E\left(\sum_{i=1}^{n} x_i \mid \mu\right) = m\mu + (n - m)\mu^t \tag{14}$$

and

$$E\left(\sum_{i=1}^{n} x_i^2 \mid \mu, \sigma^2\right) = m\sigma^2 + f(\mu^t, \sigma^t)^2, \tag{15}$$

where current estimates for the parameter would be $\theta^t = (\mu^t, \sigma^{t^2})$ The M step would then maximize the function

$$l(\theta) = \sum_{i=1}^{n} \ln(f(x_i \mid \mu^t \sigma^t)). \tag{16}$$

TABLE 2.6. BMDPAM Likelihood Estimates

Love	1.00											
Conflict	−.44	1.00										
Ambivalence	−.73	.55	1.00									
Maintenance	.48	.04	−.19	1.00								
Self-esteem	−.01	−.07	−.01	−.14	1.00							
Affect	−.15	.05	.11	−.18	.04	1.00						
Nurturance	−.15	.00	.13	−.22	.00	.54	1.00					
Energy	−.05	.11	.09	−.03	.50	.00	.07	1.00				
Unadaptive	−.09	.12	.12	−.02	.10	.10	.09	.24	1.00			
Unpredictable	−.14	−.03	.06	−.12	.07	.10	.09	.08	.11	1.00		
Dull	−.06	.09	.07	.11	.00	−.07	−.13	.05	.01	.02	1.00	
Fussy-distraught	−.12	.16	.04	−.08	.00	−.04	.07	.15	.33	.30	.32	1.00

Table 2.6 gives the BMDPAM likelihood estimates for the sample data set.

5. Longitudinal Methods

Techniques already discussed can also be used for estimating missing data in longitudinal designs. The additional dimension of time produces additional complexity for missing data patterns (e.g., due to dropouts) and provides additional data for estimating missing values (i.e., the availability of repeated measures that can act as predictors of the missing data point.) After listing a set of suggested techniques, we will demonstrate the most flexible and possibly the most useful method: likelihood estimates when the pattern of missing data can be assumed to be nested. Methods for estimating missing data in longitudinal studies along with their benefits and drawbacks are presented in Table 2.7.

5.1. Patterns of Missing Data

As in the cross-sectional data set, the desired missing data pattern has data missing completely at random. Optimally no mechanism, including knowledge of the particular wave, affects the occurrence of missing data. However, since individuals drop out of studies, any variable in a wave with dropouts will predict the occurrence of missing data in any other variable in that wave. Marini, Olsen, and Rubin (1980) have demonstrated that dropouts can be handled as long as the pattern is nested. The likelihood function can be decomposed provided that individuals do not dropout and then return to the study.

TABLE 2.7. Methods of Missing Data Estimation for Longitudinal Studies

Method	Assets	Liabilities	Pattern
Direct parameter estimation			
Listwise deletion	Generates consistent matrices	Decimates your data set	Missing completely at random (MCAR)
Pairwise analysis	Makes use of all available data	Can yield inconsistent matrices	MCAR
Maximum likelihood estimation	Makes use of maximum amount of predictive information	Hard to compute Dependent on variable selection	MAR
Mean insertion	Easy to compute	Depresses variance Often gives strange correlation and covariance matrices Can unintentionally increase the size of ANOVA effects	MAR
Subroup mean insertion	Maintains some variance	Dependent on the selection of groups and waves	MAR
Random assignment within subgroups	Maintains some variance	Dependent on the selection of groups	MAR
Estimate under different regression models	Sensitive to individual differences Allows prediction across waves	Limited number of variables	MAR

The complexity of missingness in longitudinal data becomes apparent when one considers that estimates may make use of within- or across-occasion data. One is typically interested in longitudinal questions because of hypotheses involving change over time. The choice of using within-time predictors of missing values depends upon the expectations of patterns of change and covariation over time. As in the cross-sectional case, one can estimate parameters or individual data points.

5.2. Mean and Subgroup Mean Insertion

The problem with selecting a mean for variables measured repeatedly is essentially which mean to select. In the case of data missing completely at random, the mean of respondents within a wave may seem most appropriate. Information, however, may be available for the same variable in a prior wave. In fact, in many multiwave studies, respondents sometimes miss a single administration of a questionnaire in the midst of data collection. In this case, a profile of the missing variable for the group is available along with a partial profile for the individual. These profiles use more information than would the single wave mean. Insertion of a within-wave mean amounts to the implicit assumption of low stability developmentally. Assuming a trend in the direction shown, the insertion of the within-wave (pooled) mean has the tendency to strengthen monotonic increasing or decreasing trends when the estimate occurs in waves other than the first or last while weakening trends in the latter two cases.

When the data are missing completely at random within each wave, there is no reason to believe that nonrespondents are systematically different from respondents. estimates based on trends, then, seem to be appropriate.

5.3. Insertion Using a Regression Model

Depending on the size of the autocorrelation among repeated measures of the same variable, stabilities can be used to improve the regression estimate of the missing data point. For regression, prior values of the missing data point can be included in the regression equation along with a set of other predictors.

5.4. Trend Analysis

To the degree that a general curve or trend can be established for a repeated-measures variable, a functional equation representing the longitudinal pattern can be used to estimate the missing point. The shape of the function across waves places the individual along a characteristic curve. This assumes that a typical curve suggests the trend equally well for all or at least a subgroup of respondents. This is equivalent to requiring that data be missing at random at each wave.

Kleinbaum (1973) suggested using the generalized growth curve model to handle missing data estimation in panel designs (Thissen, this volume).

For measures repeated across many occasions, it is often the case that patterns of change can be described by either a single or a few characteristic curves. Often, for example, polynomials of first-order (linear regression) or second-order (quadratic regression) will adequately describe data from four occasions. It then makes some sense to interpolate or extrapolate missing values on the basis of the characteristic curve and the consideration of which data points are available. Note, however, that first- and second-order polynomials require at least two and three observations, respectively, for the coefficients to be estimated.

The use of growth curves can be considered a two-stage process. The first step involves determining how many qualitatively different curves adequately describe the data. The second step then involves the estimation of the missing values according to the individual's characteristic curve. Kleinbaum (1973) presents an estimator for a purely linear growth curve. In the case of higher-order or nonlinear functions describing a single curve approapriate for the sample, the function can be used to estimate the missing data point. For multiple functions, however, the decision of how many functions may be appropriate is difficult. Once the decision has been made, though, methods of latent growth curve analysis (McArdle and Epstein, 1987) can be used to test the adequacy of the curve assignment.

One method that might prove useful involves typing individuals according to the patterns of residuals. For four times of measurement, the variable in question can be transformed (for complete cases only) into orthogonal polynomials (i.e., linear, quadratic, and cubic). Individuals can then be typed according to the size of the residuals on each of the transformed variables. Grouping values can then be assigned according to the characteristic curve (von Eye and Nesselroade, in press).

5.5. Direct Parameter Estimation

5.5.1. Listwise Deletion. Given how many variables are to be used in a particular analysis, the multiwave design tends to lose subjects because subjects are required to have complete data across all waves in a complete subjects analysis. Although listwise deletion always yields an invertible matrix, the analysis may not pertain to nonrespondents.

5.5.2. Pairwise Deletion. Once again, pairwise deletion is dependent on the MAR pattern to yield an analyzable matrix and to generate a representative solution. Since, a systematic mechanism (knowledge of the wave) often generates the missing data strucure, non-positive-definite matrices are more likely to result.

5.5.3. Likelihood Estimation with Nested Missing Data Patterns. Marini et al. (1980)[1] have shown that the MCAR pattern is not strictly necessary to generate unbiased and efficient estimates in panel studies. If the pattern of missing data is nested (that is, subjects missing at one wave remain missing at all followup waves), then the likelihood function can be factored by establishing all patterns of missing data in the sample. The analysis can be treated as a multiple-group analysis in which common parameters across different patterns of missing data can be constrained to be equal, yielding a set of parameters. The strategy for obtaining this analysis is similar to the methods suggested by Allison (1982) for the multiple-group cross-sectional estimation of missing data.

When the pattern of missing data is not strictly nested, Marini et al. (1980) suggest deleting those cases that do not conform to the pattern under the assumption that there is no mechanism that can account for those subjects returning to the study.

The strategy for estimating the missing data resides in the characteristics of the likelihood of the observed data. These characteristics can be shown for the bivariate case. The estimation of the likelihood for the bivariate case has been presented by Marini, Olsen, and Rubin (1980) and can be summarized as follows.

Let x and y represent variables with two different missing data structures (n observations on y and $n - m = N$ observations on x) with x measured at time 1 and y measured at time 2. These variables can be described by a set of parameters

$$\theta = (\mu_x, \mu_y, \sigma_x^2, \sigma_y^2, \sigma_y^2), \tag{17}$$

where μ_x, σ_x^2, μ_y, σ_y^2 are the respective means and variances of x and y and σ_{xy} is the covariance of the two variables.

The likelihood of observing the N (x, y) pairs is then

$$L(\theta) = \prod_{i=1}^{N} f(x_i, y_i \mid \mu_x, \mu_y, \sigma_x^2, \sigma_y^2, \sigma_{xy}). \tag{18}$$

Since x and y are dependent,

$$f(x, y) = f(x)f(y \mid x). \tag{19}$$

[1] Material from this section is reprinted with permission of Jossey-Bass Publishers, Marini, M. M., Olsen, A. R., and Rubin, D. B. "Maximum likelihood estimation in panel studies with missing data," *Sociology Methodology 1980* (1980).

The likelihood becomes

$$L(\theta) = \prod_{i=1}^{N} f(x \mid \mu_x, \sigma_x^2) f(y \mid \alpha + \beta_{yx}x, \sigma_{y \cdot x}^2), \tag{20}$$

with $\sigma_{y \cdot x}^2$ a residualized variance.

Since x is conditioned only on its own mean and standard deviation, the first part of the likelihood is a function of x alone. The conditional nature of y can be expressed as the regression of y on x, where

$$\beta_{xy} = \frac{\sigma_{xy}}{\sigma_{x,}^2} \qquad \alpha = \mu_y \mu_x \frac{\sigma_{xy}}{\sigma_x^2}, \tag{21}$$

and

$$\sigma_{y \cdot x}^2 = \sigma_y^2 - \frac{\sigma_{xy}^2}{\sigma_x^2}.$$

The likelihood can now be factored to represent the fact that for n of the cases only values of the x variable are present:

$$L(\theta) = \left[\prod_{i=1}^{n+m} f(x_i \mid \mu_x, \sigma_y^2) \right] \left[\prod_{i=1}^{n} f(y_i \mid \alpha + \beta_{yx} x_i, \sigma_{y \cdot x}^2) \right]. \tag{22}$$

By maximizing the right side of the equation, we get values of $L(\theta)$ that would be most likely to generate the observed values of x and y. These parameters can be then used to give estimates of the original parameters. The maximization can be accomplished by maximizing the function in each set of brackets separately. The factorization represented by equation (22) can only occur if the nested pattern exists, since we require that y be conditioned on x but x not be conditioned on y. This requirement is only natural, considering that y was observed after x.

For three variables x, y, and z measured respectively at three different waves of data collection, n people responded to all three waves, m people responded to the first two waves only, and k people responded to the first wave only. The likelihood for the situation is

$$L(\theta) = \prod_{i=1}^{N} f(x_i y_i z_i \mid \mu_x \mu_y \mu_z \sigma_x^2 \sigma_y^2 \sigma_z^2 \sigma_{xy} \sigma_{xz} \sigma_{yz}), \tag{23}$$

where $N = n + m + k$.

Following the same logic, we obtain the factored form of the likelihood:

$$L(\theta) = \left[\prod_{i=1}^{n+m+k} f(x_i \mid \mu_x, \sigma_x^2)\right] \cdot \left[\prod_{i=1}^{n+m} f(y_i \mid \alpha_y + \beta_{yx}x_i, \sigma_{y\cdot x}^2)\right] \cdot \left[\prod_{i=1}^{n} f(z_i \mid \alpha_z + \beta_{zx}x_i + \beta_{zy}y_i, \sigma_{z\cdot yx}^2)\right]. \quad (24)$$

This equation tells us two things. First, by getting ML estimates for the parameters μ_x, σ_x^2, σ_y^2, β_{yz}, $\sigma_{y\cdot x}^2$, α_z, β_{zx}, β_{zy}, and $\sigma_{z\cdot yx}^2$, we can generate the estimates for the means $\hat{\bar{x}}$, $\hat{\bar{y}}$, $\hat{\bar{z}}$, variances $\hat{s}_x^2$, $\hat{s}_y^2$, $\hat{s}_z^2$, and covariances $\hat{s}_{xy}$, $\hat{s}_{xz}$, $\hat{s}_{yz}$. These are the estimates we wish to obtain. Second, because of the way in which the likelihood is bracketed, we can essentially move from wave to wave in the estimation procedure. Estimating parameters in the first bracket requires only the wave 1 variable. These estimates can be "passed along" to the second bracket that only requires wave 1 estimates and the wave 2 variable. These estimates can then be passed along to the third bracket, which requires only wave 1 and wave 2 estimates and the wave 3 variable. The parameter estimates that this procedure will yield can then be used in subsequent analyses. Marini et al. (1980) extend this logic to deal not just with a single repeatedly measured variable, but with a set of variables measured at each wave.

The computational method to get the estimates is based on the following rationale. When the pattern of missing data is nested, one can partition the data set into subsets. Each subset represents those with complete data up to and including that wave. Matrices of means and covariances can be generated for each subset. Subset 1 represents all those with complete data at the first wave. since everyone has data on these variables (i.e., there is no missing data on these variables) the means and variance/covariance matrix can be estimated directly. Subset 2 includes all of those with complete data on the first and second waves. To estimate their means and variance/covariance matrix, we could just use the data in this subset. We have, however, additional data, namely, the estimates from subset 1. For the x (wave 1) variables these are more complete, since they were computed using more data. So these subset 1 estimates can be passed on to subset 2. Similarly, subset 3 estimates can use the estimates generated in subset 2. In this manner we can use all available data to generate our estimates of the means and variance/covariance matrices.

5.6. An Example of Missing Data with a Nested Longitudinal Pattern

The following example illustrates how estimation is accomplished. Descriptive statistics for three variables measured at three different occasions are presented in Table 2.8. For this analysis we selected three of the Braiker–Kelley marriage variables (LOVE, CONFlict, and AMBivalence), each of which was measured at three waves of data

TABLE 2.8. Descriptive Statistics and Correlation Matrices for Braiker–Kelly Subscales

Variable			Mean	Standard deviation	N
Wave 1	X_1	W3LOVE	78.04	9.14	158
	X_2	W3CONF	20.73	6.33	158
	X_3	W3AMB	10.56	5.52	158
Wave 2	Y_1	W9LOVE	76.49	9.56	162
	Y_2	W9CONF	21.69	6.02	162
	Y_3	W9AMB	11.72	5.91	162
Wave 3	Z_1	XWLOVE	74.30	12.83	131
	Z_2	XWCONF	22.46	6.89	131
	Z_3	XWAMB	12.60	7.46	131

		1.00	−.43	−.72						
Wave 1	**A** =	−.43	1.00	.55						
		−.72	.55	1.00						
		1.00	−.42	−.71	−.77	−.36	−.40			
		−.42	1.00	.52	−.30	.64	.34			
Waves 1, 2	**B** =	−.71	.52	1.00	−.57	.41	.62			
		.77	−.30	−.57	1.00	−.39	−.60			
		−.36	.64	.41	−.39	1.00	.53			
		−.40	.34	.62	−.60	.53	1.00			
		1.00	−.42	−.68	.74	−30	−37	.42	−.17	−.24
		−.42	1.00	.50	−.33	.65	.35	−.31	.60	.33
		−.68	.50	1.00	−.59	.39	.64	−.36	.29	.42
		.74	−.33	−.59	1.00	−.36	−.64	.53	−.28	−.42
Waves 1, 2, 3	**C** =	−.30	.65	.39	−.36	1.00	.52	−.28	.60	.38
		−.37	.35	.64	−.64	.52	1.00	−.38	.33	.57
		.42	−.31	−.36	.53	−.28	−.38	1.00	−.50	−.79
		−.17	.60	.29	−.28	.60	.33	−.50	1.00	.60
		−.24	.33	.42	−.42	.38	.57	−.79	.60	1.00

collections. The pattern of missing data is nested, which means essentially that individuals dropping out do not return to the study. The strategy for estimating a single correlation or covariance matrix for all nine variables can thus use the fact that parameter estimates for subgroup patterns suggested above depend on data from that wave and prior waves for a particular subgroup and not on any data from future waves. To allow us to use matrix notation, we will refer to the wave 1 variables as X_1, X_2, and X_3; wave 2 as Y_1, Y_2, and Y_3; and wave 3 as Z_1, Z_2, and Z_3.

The first step in the analysis is to calculate a vector of means and a variance/covariance matrix on all present data in the first wave. To make the analysis somewhat easier to comprehend, we will use a correlation matrix and standardized variables (mean = 0, standard deviation = 1). The technique will apply to covariance matrices and nonstandardized variables. A matrix is calculated for the first-wave variables using listwise deletion. Because of the nested pattern, the number of cases for this matrix equals the number of people enrolled in the study. This matrix, **A**, and subsequent matrices for this analysis appear in Table 2.8. To generate the matrix, a constant (similar to the constant used in a regression design matrix) has been placed in the **A**(1, 1) position. This matrix is then "swept" first on the constant and then on the wave 1 variables. The sweep operator has been defined by Dempster (1969) as follows:[2]

> A $p \times p$ matrix, **M**, will be said to have been *swept on row k and column k* if **M** is replaced by another $p \times p$ matrix **N** whose (i, j) elements n_{ij} are related to the (i, j) elements m_{ij} of **M** as follows:
>
> $$n_{kk} = -1/m_{kk}$$
> $$n_{ik} = m_{ik}/m_{kk}$$
> $$n_{kj} = m_{kj}/m_{kk}$$
> $$n_{ij} = m_{ij} - m_{ik}m_{kj}/m_{kk} \qquad \text{for } i \neq k \text{ and } j \neq k.$$

Sweeping on the constant creates a new matrix, **ASWEPTC**, in which (1) the first row formerly of means has been replaced by coefficients representing the regression of the original variables onto the constant, and (2) the original variance/covariance matrix is replaced by the residual covariance matrix from this regression. Sweeping on the remaining variables creates a matrix **ASWEPT**, which can be passed onto the next step.

[2] Reprinted with permission from Addison-Wesley, Dempster, A. P. *Elements of Continuous Multivariate Analysis* (1969).

Next, the variables measured at wave 2 are regressed onto the variables measured at wave 1. To accomplish this, we create a vector of means and a variance/covariance matrix of the wave 1 and wave 2 variables using listwise deletion for those variables. These are combined into a matrix **B**, with a constant once again in the **B**(1, 1) position. This matrix is then swept on the constant and the wave 1 variables. Sweeping a set of variables $\{1, X_1, X_2, X_3, Y_1, Y_2, Y_3\}$ on the constant and the X variables is the equivalent of regressing the Y variables onto the X variables. Since **B** can be partitioned as

$$\mathbf{B} = \begin{bmatrix} \mathbf{X'X} & \mathbf{X'Y} \\ \mathbf{Y'X} & \mathbf{Y'Y} \end{bmatrix}, \tag{25}$$

then

$$\mathbf{BSWEPT} = \begin{bmatrix} (\mathbf{X'X})^{-1} & (\mathbf{X'X})^{-1}\mathbf{X'Y} \\ -\mathbf{Y'X}(\mathbf{X'X})^{-1} & \mathbf{Y'}(\mathbf{I}-\mathbf{X}(\mathbf{X'X})^{-1}\mathbf{X'})\mathbf{Y} \end{bmatrix} = \begin{bmatrix} \mathbf{B1} & \mathbf{B2'} \\ \mathbf{B2} & \mathbf{B3} \end{bmatrix}, \tag{26}$$

in which the off-diagonal submatrices are the regression weights and the lower right submatrix is the residual variance/covariance matrix of the regression. We next create a new matrix, **BCAT**, as follows:

$$\mathbf{BCAT} = \begin{bmatrix} \mathbf{ASWEPT} & \mathbf{B2'} \\ \mathbf{B2} & \mathbf{B3} \end{bmatrix}. \tag{27}$$

With another wave of data to consider, we sweep this matrix on the Y variables to create the matrix **BCATSWPY**, which will be passed onto the next step.

Next, we create a vector of means and a variance/covariance matrix of the X, Y, and Z variables, using listwise deletion. The number of cases for this matrix will reflect all those who completed all three waves. These are combined into a matrix **C**, which is then swept on the constant and the X and Y variables to form **CSWEPT**, which can be partitioned into

$$\mathbf{CSWEPT} = \begin{bmatrix} \mathbf{C1} & \mathbf{C2'} \\ \mathbf{C2} & \mathbf{C3} \end{bmatrix}, \tag{28}$$

in which **C2** is the matrix of weights for the regression of the **Z** variables onto the X and Y variables and **C3** is the residual variance/covariance for that regression. This matrix is combined with **BCATSWPY** to form the matrix

$$\mathbf{CCAT} = \begin{bmatrix} \mathbf{BCATSWPY} & \mathbf{C2'} \\ \mathbf{C2} & \mathbf{C3} \end{bmatrix}. \tag{29}$$

The final step requires that we reverse-sweep **CCAT** on the X and Y

variables to obtain ML estimates. The reverse sweep is defined by Dempster (1969) as

$$\mathbf{M} = \mathbf{REVSWP(N)} \qquad \text{when } m_{kk} = \frac{-1}{n_{kk}}$$

$$m_{ik} = \frac{-n_{ik}}{n_{kk}}$$

$$m_{kj} = \frac{-n_{kj}}{n_{kk}}$$

$$m_{ij} = n_{ij} - \frac{n_{ik}n_{kj}}{n_{kk}} \qquad \text{for } j \neq i \text{ and } k \neq i.$$

Reverse–sweeping **CCAT** yields the matrix **MAXEST**, in which the first row (or column) contains the estimates for the means, and the remainder of the matrix contains the ML missing data estimates of the variance/covariance matrix. The solution matrix for our sample data appears in Table 2.9. The listing of the SAS PROCIML program to compute the estimates appears in the appendix. Had we only two waves of data, we could have stopped the process after computing **BCAT** and used a reverse sweep at that point on the X variables only to calculate estimates for the means and variance/covariance matrix based on those two waves of data.

Comparing the estimates in Table 2.9 with the input matrices suggests that the results of this analysis yields estimates that fall within bounds defined by the wave 1, 2, and 3 correlations. This result is satisfying in that the procedure appears to combine and weight the data in a way that yields intuitively reasonable results.

TABLE 2.9. Longitudinal Maximum Likelihood Estimates for Correlation Matrix

W3LOVE	1.00								
W3CONF	−.42	1.00							
W3AMB	.71	.52	1.00						
W9LOVE	.77	−.30	−.57	1.00					
W9CONF	−.36	.64	.41	−.14	1.00				
W9AMB	−.40	.34	.62	−.18	.53	1.00			
XWLOVE	.43	−.29	−.35	−.22	−.29	−.36	.99		
XWCONF	−.19	.59	.29	−.03	.58	.31	−.49	.98	
XWAMB	−.26	.32	.41	−.02	.38	.55	−.78	.58	.98

In the case in which the pattern of missing data is not nested, methods are available for estimating parameters (Little and Rubin, 1987). If only a few cases do not conform to the pattern, it may make sense to delete those cases in order to avoid the additional complexity.

6. Conclusion

This chapter attempted to discuss many of the currently used methods for estimating missing data in developmental data sets. Although we believe that each data set presents somewhat unique problems for analysis, we do recommend certain general procedures. The methods of ML estimation (EM algorithms, multiple-group direct parameter estimation, and longitudinal estimation with a nested pattern) appear to be viable alternatives to more widely used methods, such as mean imputation and listwise deletion. Comparing these methods with less computer-intensive methods shows that these methods seem to have the advantages of both, namely, maximal use of data and well-behaved estimates. Because many of these methods are widely available in computer packages (BMDP, LISREL) and others are easily programmed, they can be used to create a matrix or a data set that yields estimates that more accurately reflect the data collected.

Appendix

A.1. LISREL Program for Cross-Sectional Maximum-Likelihood Missing Data Estimation

```
INPUT PROGRAM
NUMERIC A
END FILE
END INPUT PROGRAM
USERPROC NAME = LISREL
TIELE ALLISON ESTIMATION (GROUP 1 COMPLETE DATA GROUP)
DA NG = 2 NI = 5 NOBS = 155 MA = KM
LABELS
*
  'W3FUSDIS'  'W3DULL'  'W3LOVE'  'W3UNADAP'  'W3UNPRED'
```

```
KM FU
*
    1.00000   0.34777  -0.11592   0.30890   0.27060
    0.34777   1.00000  -0.06526   0.04105   0.02977
   -0.11592  -0.06526   1.00000  -0.09368  -0.13883
    0.30890   0.04105  -0.09368   1.00000   0.09044
    0.27060   0.02977  -0.13883   0.09044   1.00000
SE
  'W3FUSDIS' 'W3DULL' 'W3LOVE' 'W3UNADAP' 'W3UNPRED'/
MO        NY = 5    NE = 5            LY = FU,FI  C
    BE = FU,FI
              TE = SY,FI    PS = SY,FI
PA LY
*
0 0 0 0 0
0 0 0 0 0
0 0 0 0 0
0 0 0 0 0
0 0 0 0 0
PA TE
*
0
0 0
0 0 0
0 0 0 0
0 0 0 0 0
PA BE
*
0 0 0 0 0
0 0 0 0 0
1 0 0 0 0
1 1 0 0 0
0 0 1 0 0
PA PS
*
0
0 0
0 0 1
0 0 0 1
0 0 0 0 1
MA LY
*
1.0   0   0   0   0
```

```
 0     1.0    0    0    0
 0       0  1.0    0    0
 0       0    0  1.0    0
 0       0    0    0  1.0
MA TE
*
0
0  0
0  0  0
0  0  0  0
0  0  0  0  0
MA BE
*
0  0  0  0  0
0  0  0  0  0
0  0  0  0  0
0  0  0  0  0
0  0  0  0  0
MA PS
*
1.0
 0     1.0
 0       0   .5
 0       0    0   .5
 0       0    0    0   .5
OU PT SE TV NS MI SS RS TM = 40
TITLE ALLISON ESTIMATION (GROUP 2 MISSING DATA GROUP)
DA NG = 2 NI = 5 NOBS = 15 MA = KM
LABELS
*
   'W3FUSDIS'  'W3DULL'  'W3LOVE'  'W3UNADAP'  'W3UNPRED'
KM FU
*
    1.00000    0.34777   -0.11592    0.30890    0.27060
    0.34777    1.00000   -0.06526    0.04105    0.02977
   -0.11592   -0.06526    1.00000   -0.09368   -0.13883
    0.30890    0.04105   -0.09368    1.00000    0.09044
    0.27060    0.02977   -0.13883    0.09044    1.00000
SE
  'W3FUSDIS'  'W3DULL'  'W3LOVE'  'W3UNADAP'  'W3UNPRED'/
M0          NY = 5      NE = 5                LY = FU,FI   C
     BE = IN         TE = FU,FI         PS = IN
EQ LY(1, 1, 1) LY(1, 1)
```

```
EQ LY(1, 2, 2) LY(2, 2)
EQ LY(1, 4, 4) LY(4, 4)
EQ LY(1, 5, 5) LY(5, 5)
FI LY(3, 3)
EQ TE(1, 1, 1) TE(1, 1)
EQ TE(1, 1, 1) TE(1, 1)
EQ TE(1, 2, 2) TE(2, 2)
EQ TE(1, 4, 4) TE(4, 4)
EQ TE(1, 5, 5) TE(5, 5)
FI TE(3, 3)
ST 0.0 LY(3, 3)
ST 1.0 TE(3, 3)
OU PT SE TV NS MI SS RS TM = 40
END USER
```

A.2. SAS IML Program for Longitudinal Maximum-Likelihood Missing Data Estimation

```
PROC IML;
START;
CONSTANT = 1;
***                                                        ;
*** WAVE 1 MEANS AND VARIANCE/COVARIANCE MATRIX;
***                                                        ;
INDEX = {1};
MEANS1 = {0.0   0.0   0.0};
COV1 = {1.00   -.43   -.72,
        -.43   1.00    .55,
        -.72    .55   1.00};
***                                    ;
*** CONCATENATE WAVE 1 MATRIX;
***                                    ;
A1 = CONSTANT||MEANS1;
A2 = MEANS1'||COV1;
A = A1//A2;
PRINT 1;
CCAT = A;
***                               ;
```

```
*** SWEEP WAVE 1 MATRIX;
***                         ;
ASWEPT = {0  0  0  0,
          0  0  0  0,
          0  0  0  0,
          0  0  0  0};
ASWEEPST = 1;
ASWEEPEN = 4;
ROWS = 4;
COLUMNS = 4;
DO K = ASWWEPST TO ASWEEPEN;
   ASWEPT(|K, K|) = −1/CCAT(|K, K|);
DO I = 1 TO ROWS;
   IF I = K THEN GOTO CONTIN2;
   ASWEPT(|I, K|) = CCAT(|I, K|)/CCAT(|K, K|);
   CONTIN2: COMMENT;
DO J = 1 TO COLUMNS;
   IF J = K THEN GOTO CONTIN3;
   ASWEPT(|K, J|) = CCAT(|K, J|)/CCAT(|K, K|);
   CONTIN3:  COMMENT;
   IF I = K THEN GOTO CONTIN1;
   IF J = K THEN GOTO CONTIN1;
ASWEPT(|I, J|) = CCAT(|I, J|) − CCAT(|I, K|)#CCAT(|K, J|)/CCAT(|K, K|);
  CONTIN1:  COMMENT;
END;
END;
PRINT ASWEPT K "RUN";
CCAT = ASWEPT;
END;
PRINT ASWEPT;
CCAT = ASWEPT;
***                                                  ;
*** WAVE 2 MEANS AND VARIANCE/COVARIANCE MATRIX;
***                                                  ;
MEANS2 = {0.0  0.0  0.0  0.0  0.0  0.0}
COV2 = {1.00  −.42  −.71  −.77  −.36  −.40,
        −.42  1.00   .52  −.30   .64   .34,
        −.71   .52  1.00  −.57   .41   .62,
         .77  −.30  −.57  1.00  −.39  −.60,
        −.36   .64   .41  −.39  1.00   .53,
        −.40   .34   .62  −.60   .53  1.00};
***                                ;
*** CONCATENATE WAVE 2 MATRIX;
```

```
***                                    ;
B1 = CONSTANT||MEANS2;
B2 = MEANS'||COV2;
B = B1//B2;
PRINT B;
INDEX2 = {1  2  3  4}:
CCAT = B:
***                          ;
*** SWEEP WAVE 2 MATRIX;
***                          ;
BSWEPT = {0  0  0  0  0  0  0,
          0  0  0  0  0  0  0,
          0  0  0  0  0  0  0,
          0  0  0  0  0  0  0,
          0  0  0  0  0  0  0,
          0  0  0  0  0  0  0,
          0  0  0  0  0  0  0}
BSWEEPST = 2;
BSWEEPEN = 7;
ROWS = 7;
COLUMNS = 7;
DO K = BSWEEPST TO BSWEEPEN;
   BSWEPT(|K, K|) = -1/CCAT(|K, K|);
DO I = 1 TO ROWS;
   IF I = K THEN GOTO CONTIN4;
   BSWEPT = (|I, K|) = CCAT(|I, K|)/CCAT(|K, K|);
   CONTIN4: COMMENT;
DO J = 1 TO COLUMNS;
   IF J = K THEN GOTO CONTIN5;
   BSWEPT = (|K, J|) = CCAT(|K, J|)/CCAT(|K, K|);
   CONTIN5: COMMENT;
   IF I = K THEN GOTO CONTIN6;
   IF J = K THEN GOTO CONTIN6;
BSWEPT(|I, J|) = CCAT(|I, J|) - CCAT(|I, K|)#CCAT(|K, J|)/CCAT(|K, K|);
   CONTIN6: COMMENT;
END;
END;
CCAT = BSWEPT;
END;
PRINT BSWEPT "RUN";
CCAT = BSWEPT;
PRINT CCAT;
PRINT CCAT;
```

```
BCAT21 = BSWEPT(|{5 6 7}, {1 2 3 4}|);
BCAT12 = BCAT21';
BCAT22 = BSWEPT(|{5 6 7}, {5 6 7}|);
INDEX = {2 3 4};
BCAT1 = ASWEPT||BCAT12;
BCAT2 = BCAT21 ||BCAT22;
BCAT = BCAT1//BCAT2;
PRINT BCAT21;
PRINT BCAT12;
PRINT BCAT22;
PRINT BCAT;
***                                                  ;
*** WAVE 3 MEANS AND VARIANCE/COVARIANCE MATRIX;
***                                                  ;
MEANS3 = {0.0  0.0  0.0  0.0  0.0  0.0  0.0  0.0  0.0};
COV3 = {1.00  -.42  -.68   .74  -.30  -.37   .42  -.17  -.24,
        -.42  1.00   .50  -.33   .65   .35  -.31   .60   .33,
        -.68   .50  1.00  -.59   .39   .64  -.36   .29   .42,
         .74  -.33  -.59  1.00  -.36  -.64   .53  -.28  -.42,
        -.30   .65   .39  -.36  1.00   .52  -.28   .60   .38,
        -.37   .35   .64  -.64   .52  1.00  -.38   .33   .57,
         .42  -.31  -.36   .53  -.28  -.38  1.00  -.50  -.79,
        -.17   .60   .29  -.28   .60   .33  -.50  1.00   .60,
        -.24   .33   .42  -.42   .38   .57  -.79   .60  1.00};
***                              ;
*** CONCATENATE WAVE 3 MATRIX;
***                              ;
C1 = CONSTANT||MEANS2;
C2 = MEANS3'| |COV3;
C = C1/C2;
CCAT = C;
PRINT C;
INDEX3 = {1 2 3 4 5 6 7};
***                      ;
*** SWEEP WAVE 3 MATRIX;
***                      ;
CSWEPT = {0  0  0  0  0  0  0  0  0  0,
          0  0  0  0  0  0  0  0  0  0,
          0  0  0  0  0  0  0  0  0  0,
          0  0  0  0  0  0  0  0  0  0,
          0  0  0  0  0  0  0  0  0  0,
          0  0  0  0  0  0  0  0  0  0,
          0  0  0  0  0  0  0  0  0  0,
```

```
            0 0 0 0 0 0 0 0 0 0,
            0 0 0 0 0 0 0 0 0 0,
            0 0 0 0 0 0 0 0 0 0}:
CSWEEPST = 1;
CSWEEPEN = 7;
ROWS = 10;
COLUMNS = 10;
DO K = CSWEEPST TO CSWEEPEN;
    CSWEPT(|K, K|) = −1/CCAT(|K, K|);
DO I = 1 TO ROWS;
    IF I = K THEN GOTO CONTIN7;
    CSWEPT = (|I, K|) = CCAT(|I, K|)/CCAT(|K, K|);
    CONTIN7: COMMENT;
DO J = 1 TO COLUMNS;
    IF J = K THEN GOTO CONTIN8;
    CSWEPT = (|K, J|) = CCAT(|K, J|)/CCAT(|K, K|);
    CONTIN8: COMMENT;
    IF I = K THEN GOTO CONTIN9;
    IF J = K THEN GOTO CONTIN9;
CSWEPT(|I, J|) = CCAT(|I, J|) − CCAT(|I, K|)#CCAT(|K, J|)/CCAT(|K, K|);
    CONTIN9: COMMENT;
END;
END;
PRINT CSWEPT "RUN";
CCAT = CSWEPT;
END;
PRINT CSWEPT;
CCAT = CSWEPT;
CCAT21 = CSWEPT(|{8 9 10}, {1 2 3 4 5 6 7}|);
CCAT12 = CCAT21′;
CCAT22 = CSWEPT(|{8 9 10}, {8 9 10}|);
INDEXB2 = {5 6 7};
CCAT11 = BSWEPT;
PRINT CCAT11;
PRINT CCAT12;
PRINT CCAT21;
PRINT CCAT22;
CCAT1 = CCAT11||CCAT12;
CCAT2 = CCAT21||CCAT22;
CCAT = CCAT1//CCAT2;
PRINT CCAT;
INDEX4 = {2 3 4 5 6 7};
***                                  ;
```

```
*** REVERSE SWEEP FINAL MATRIX;
***                            ;
REVSWP = {0 0 0 0 0 0 0 0 0 0
          0 0 0 0 0 0 0 0 0 0
          0 0 0 0 0 0 0 0 0 0
          0 0 0 0 0 0 0 0 0 0
          0 0 0 0 0 0 0 0 0 0
          0 0 0 0 0 0 0 0 0 0
          0 0 0 0 0 0 0 0 0 0
          0 0 0 0 0 0 0 0 0 0
          0 0 0 0 0 0 0 0 0 0
          0 0 0 0 0 0 0 0 0 0};
REVSWPST = 2;
REVSWPEN = 7;
ROWS = 10;
COLUMNS = 10;
DO K = REVSWPST TO REVSWPEN;
   REVSWP(|K, K|) = 1/CCAT(|K, K|);
   REVSWP(|K, K|) = -REVSWP(|K, K|);
DO I = 1 TO ROWS;
   IF I = K THEN GOTO CONTIN10;
   REVSWP = (|I, K|) = -CCAT(|I, K|)/CCAT(|K, K|);
   CONTIN10: COMMENT;
DO J = 1 TO COLUMNS;
   IF J = K THEN GOTO CONTIN8;
   REVSWP = (|K, J|) = -CCAT(|K, J|)/CCAT(|K, K|);
   CONTIN11: COMMENT;
   IF I = K THEN GOTO CONTIN12;
   IF J = K THEN GOTO CONTIN12;
REVSWP(|I, J|) = CCAT(|I, J|) - CCAT(|I, K|)#CCAT(|K, J|)/CCAT(|K, K|);
   CONTIN12: COMMENT;
END;
END;
CCAT = REVSWP;
END;
PRINT REVSWP;
***                                  ;
*** PRINT MAXIMUM-LIKELIHOOD ESTIMATES;
***                                  ;
MAXEST = RECSWP;
PRINT MAXEST;
FINISH;
RUN;
```

References

Afifi, A. A., and Elashoff, R. M. (1966). "Missing observations in multivariate statistics. I: Review of the literature," *Journal of the American Statistical Association.* **61,** 595–604.

Allison, P. D. (1982). "Maximum likelihood estimation of linear models when data are missing." Unpublished manuscript, University of Pennsylvania, Philadelphia.

Anderson, J. C., and Gerbing, P. W. (1988). "Structural equation modeling in practice: A review and recommended two-step aproach," *Psychological Bulletin.* **103,** 411–423.

Anderson, J. C., and Gerbing, P. W. (1984). "The effect of sampling error on convergence, improper solutions and goodness-of-fit indices for maximum-likelihood confirmatory factor analysis," *Psychometrika.* **49,** 155–173.

Belsky, J., Gilstrap, B., and Rovine, M. (1984). "The Pennsylvania Infant and Family Development Project. I: Stability and change in mother-infant and father-infant interaction in a family setting at 1-, 3-, and 9-months," *Child Development.* **55,** 692–705.

BMDP Statistical Software (1985). Berkeley, CA: University of California Press.

Buck, S. F. (1960). "A method of estimation of missing values in multivariate data suitable for use with an electronic computer," *Journal of the Royal Statistical Society, B.* **22,** 302–306.

Cohen, J., and Cohen, P. (1975). *Applied multiple regression/correlation analysis for the behavioral sciences.* Hillsdale, N.J.: Lawrence Erlbaum.

Dempster, A. P. (1969). Elements of Continuous Multivariate Analysis. Reading, Mass: Addison-Wesley.

Dixon, W. J. (Ed.) (1983). *BMDP statistical software,* revised printing. Berkeley, Calif.: University of California Press.

Glasser, M. (1964). "Linear regression analysis with missing observations among the independent variables," *Journal of the American Statistical Association.* **59,** 834–844.

Greenlees, W. S., Reece, J. S., and Zieschang, K. D. (1982). Imputation of missing values when the probability of response depends on the variable being imputed," *Journal of the American Statistical Association.* **77,** 251–261.

Haitovsky, Y. (1968). "Missing data in regression analysis," *Journal of the Royal Statistical Society B.* **30,** 67–81.

Joreskog, K. G., and Sorbom, D. (1984). *LISREL VI: User's guide.* Mooresville, IN Scientific Software.

Kim, J. O., and Curry, J. (1977). "The treatment of missing data in multivariate analysis," *Sociological Methods and Research.* **6,** 215–240.

Kleinbaum, D. G. (1973). "A generalization of the growth curve model which allows missing data," *Journal of Multivariate Analysis.* **3,** 117–124.

Little, R. J. A., and Rubin, D. B. (1987). *Statistical analysis with missing data.* New York: Wiley.

Marini, M. M., Olsen, A. R., and Rubin, D. B. (1980). "Maximum likelihood estimation in panel studies with missing data," *Sociological Methodology* 1980, San Francisco: Jossey Bass.

McArdle, K., and Epstein, D. (1987). "Latent growth curves within developmental structural equation models," *Child Development.* **58,** 110–133.

Nie, N. H., Hull, C. H., Jenkins, J. G., Steinbrenner, K., and Bent, D. H. (1975). *SPSSx user's guide.* New York: McGraw-Hill.

Nunnally, J. C. (1978). *Psychometric theory.* New York: McGraw-Hill.

Orchard, T., and Woodbury, M. A. (1972). "A missing information principle: Theory and applications," *Proc. 6th Berkeley Symposium on Mathematical Statistics and Probability,* vol. 1, 697–715.

Rao, C. R., and Mitra, S. R. (1971). *Generalized inverse of matrices and its applications.* New York: Wiley.

Rubin, D. B. (1976). "Inference and missing data," *Biometrika.* **63,** 581–592.

SAS Institute, Inc. (1985). *SAS user's guide: Statistice,* Version 5 Edition. Cary, N.C.: SAS Institute Inc.

Timm, N. H. (1970). "The estimation of variance-covariance and correlation matrices from incomplete data," *Psychometrica.* **35**(4), 417–437.

von Eye, A., and Nesselroade, J. R. (1989, in press). Types of Change: Application of configural frequency analysis in repeated measurement designs. *Experimental Aging Research.*

Woolson, R. F., and Clarke, W. R. (1984). "Analysis of categorical incomplete longitudinal data," *Journal of the Royal Statistical Society A.* **147,** 87–99.

Chapter 3

Alternative Analyses [illegible] Repeated-Measure Designs by ANOVA and MANOVA

Paul A. Games

College of Health & Human Development
The Pennsylvania State University
University Park, Pennsylvania

Abstract

An analysis of repeated-measures data can often be done by several different methods. We start with a simple data set and illustrate several procedures that could be used, giving the virtues and limitations of each. These same considerations could be used on more complex designs. Then we go to a more complex data set with more than two levels of the repeated-measures factor. In this case, the within-subject factors can be treated either by a univariate analysis or by a multivariate analysis. To obtain an accurate univariate analysis, it is necessary to compute an index of sphericity (or circularity), usually known as epsilon, and then to adjust the df of the *MS* ratio using that index. The multivariate analysis does not require this index, but may not be very robust to leptokurtic violations of the assumption of multivariate normality. We see how to use SAS PROC GLM with the REPEATED command for the ANOVA and MANOVA, but find a major flaw in SAS when this same command is used with covariance. Finally, we look at a set of repeated-measures data where three different criteria are available so that a multivariate ANOVA and COVAR are needed to control the Type I error rate for the set of three variables.

After you mastered analysis of variance (ANOVA), you often have a wide choice in the number of possible designs available to handle a given set of data. Usually the choice of designs should be made prior to data collection so that this will dictate the possible analyses. However, often several different analyses can be done on a given data set. We shall illustrate several analyses, first on a simple set of data with only two levels for the repeated-measure factor, and then a more complex design with four levels of the repeated-measure factor, which makes a multivariate ANOVA (MANOVA) solution feasible. We hope that this will help the reader recognize the similarities and differences of various analyses.

1. Teaching Method Experiment with Pretest and Posttest

Assume we have three different experimental treatments we wish to test for teaching vector properties. *A*1 is a purely verbal method, *A*2 uses a symbolic presentation alone, and *A*3 mixes verbal components and symbolic components. Ten subjects are randomly assigned to each group. All treatments take about the same study time. All subjects take a criterion test before they start the study, and all take the same test after they finish the study. We can also compute the gain score for each subject by subtracting the pretest from the posttest score. The experimenter uses the .01 level of significance. The data are presented in Table 3.1.

1. The simplest way to analyze the data is to do a simple randomized design on the posttest scores, ignoring the pretest and gain scores. Using the nesting and crossing notation of Games (1979), we represent this as

TABLE 3.1. Data of the First Example

Treatment 1 (*A*1)			Treatment 2 (*A*2)			Treatment 3 (*A*3)			
Pre	Post	Gain	Pre	Post	Gain	Pre	Post	Gain	
8	32	24	7	27	20	6	24	18	
6	34	28	9	33	24	9	26	17	
9	36	27	6	29	23	6	24	18	
7	32	25	9	33	24	9	28	19	
5	30	25	7	28	21	5	22	17	
3	28	25	2	24	22	4	21	17	
2	29	27	4	28	24	0	20	20	
4	29	25	4	26	22	2	25	23	
0	28	28	4	28	24	0	16	16	
4	32	28	1	21	20	2	19	17	
4.8	31.0	26.2	5.3	27.7	22.4	4.3	22.5	18.2	means

an RS_{10} in (A_3) design. The three sample means are 31.0, 27.7, and 22.5, respectively, for $A1$, $A2$, and $A3$. Then

$$F = \frac{MS_A}{MS_{RS(A)}} = \frac{183.63}{11.13} = 16.49,$$

which has a probability less than .001 if the null hypothesis is true. Thus we reject the null hypothesis that the three population means are equal, and we use a Tukey WSD test (Games, 1971) to determine how they differ. The WSD stands for the wholly significant difference and controls the familywise risk of Type I error at the level of significance (here .01):

$$\text{WSD} = q_{.01,3,27}\sqrt{\frac{MS_{RS(A)}}{10}} = (4.54)(1.054) = 4.785.$$

Thus $\bar{Y}_3$ is significantly less than the other two means, but $\bar{Y}_1$ and $\bar{Y}_2$ are not significantly different. Using Duncan's (1955) underlining procedure to mark under means where the equality hypothesis is retained, we have 22.5 27.7 31.0.

Although the preceding analysis is simple, like many designs where subjects are assigned at random to treatments, and the subject variation yields the error term, it has a relatively large error term $[MS_{RS(A)} = 11.13]$ and lacks power. Since we have the pretest scores available, if they correlate positively with the posttest, we should have greater power by using the gain scores. First, however, you should check that the pretest means do not differ by more than chance. Since subjects were randomly assigned to the three treatments, they should not differ in the pretest; a test of this hypothesis yields $F < 1$. This is encouraging, since a rejection of the null here would represent a considerable threat to the interpretation of the gain scores, the posttest scores, or the analysis of covariance.

2. The ANOVA on the gain scores yields $F = MS_A/MS_{RS(A)} = 160.13/3.10 = 51.72$. Note that the error term has been reduced from 11.13 on the posttest alone to 3.10 on the gain scores. This is due to the high correlation between pretest and posttest in this data ($r = .85$). MS_A has also changed slightly, but there should be no systematic difference; however, a reduction in the denominator is expected whenever there is a positive correlation between the pretest and the posttest. The mean gains, in order of $A3$, $A2$, and $A1$ are 18.2, 22.4, and 26.2, and

$$\text{WSD} = q_{.01,3,27}\sqrt{\frac{MS_{RS(A)}}{10}} = 4.54\sqrt{\frac{3.10}{10}} = 2.5278.$$

The reduction in the error term has reduced the WSD value also, so all means are significantly different. If the correlation were lower, say a .3 or so, the reduction of the error term and the increase in power would not be as great.

3. An alternative method for using the pretest scores is to use the analysis of covariance with the posttest scores as the criterion Y and the pretest scores as the covariate X. Covariance can be conceived as an ANOVA on a criterion of the subject deviations from their predicted value, $\hat{Y}$. As in the gain score analysis, it is important that the difference between the pretest means be trivial (Elashoff, 1969; Games, 1976). The analysis of covariance assumes homogeneity of the slopes of the pretest-posttest regression in the three groups. Usually this should be tested, but many computer programs for the analysis of covariance fail to test for homogeneity of slopes. We will illustrate how to test for homogeneous slopes later. In the present data, we retain the hypothesis of homogeneous slopes on the data of Table 3.1 ($F = .94$, $p > .05$). The pooled regression slope is .96. In this case, it is obvious that the gain score analysis is very similar to the covariance analysis. In the gain score analysis, the criterion is $G = Y - 1$ (pretest score). In the covariance analysis, the criterion is $Y_{\text{residual}} = Y - .96$ (pretest score). Thus very similar results are obtained. The covariance $F = 160.04/3.20 = 50.01$, and the adjusted means are 23.0, 27.2, and 31.0 for $A3$, $A2$, and $A1$, respectively. All the adjusted means are significantly different. The use of a WSD on these adjusted means is nontrivial (and will not be covered here), since these adjusted means are no longer independent. They are no longer independent because each is computed as a "predicted value" using the equation

$$\text{Adjusted } \bar{Y}_j = \bar{Y}_j - b[\bar{X}_j - \bar{X}_{..}]$$

or a related equation expressed in matrix terms. The use of the common regression estimate of b (.96 in this example), which is solved for by pooling the variance-covariance matrices of the three groups, plus the use of the grand mean $\bar{X}_{..}$, make the adjusted means dependent. In general, covariance results are considerably more complex than ANOVA results.

The great similarity of the gain score F and the covariance F is due to the similarity of the gain score logical regression coefficient of 1.0 and the covariance regression slope of .96. With a lower correlation between the pretest and posttest, the gain score analysis and the covariance analysis would differ by more than in the present example. In this case, the treatment effects are indeed additive to the pretest scores, but in many other cases treatments might cause heterogeneous slopes and variances.

Note that the covariance means are similar to the posttest means of analysis 1, but the differences between the covariance means are similar to the differences in the gain scores of analysis 2.

Sometimes people have computed gains scores and then used the pretest as a covariate to compute a covariance. This is a waste of time and merely adds an extra layer of confusion. If it is done, you will also obtain an $F = 50.01$ as in item 3 above, and will obtain "adjusted mean gains" of 26.2, 22.4, 18.2 (as in the original mean gains). In this case, the regression slope can be shown to be $.96 - 1.0 = -.04$, which is so close to 0.0 that it has no appreciable effect. In general, the slope for such an analysis would be $b_{g.1} = b_{2.1} - 1.0$. Of course, it will not always be so close to zero, but it should be obvious that a covariance of the gain scores by the pretest used to define those prescores is an unwise (although legitimate) analysis.

4. A way to handle the present data set without getting into the complexities of covariance is to include both the pretest and posttest scores as two levels of a repeated-measures (within-subjects) factor. In this case, you are using each subject as his or her own control without having to get into multivariate or covariance analyses. When there are only two levels of the repeated-measures factor, the two-factor univariate analysis has very few assumptions. This design may be represented as an $[RS_{10}$ in $(A_3)]J_2$ design (Games, 1979). Subjects are nested under the A-factor but crossed with the J-factor. Note that the crossing operation is distributive like the mathematical operation of multiplication. In the present design, the J-factor is outside the [ ] terms and is distributed to both terms inside. Thus the J-factor is crossed with both the treatment factor (A) and the subject factor (RS). Myers (1979) calls this a mixed design because subjects are nested under one factor (A) and crossed with one factor (J). Unfortunately, this terminology is not consistent with the usual mathematical statistics terminology for a mixed design.

One of the virtues of this analysis is that it makes overt many things that may remain covert when gain scores or covariance is used. We will get six cell means, thus showing pretest means as well as posttest means:

	$J1$	$J2$
$A1$	4.8	31.0
$A2$	5.3	27.7
$A3$	4.3	22.5

In this case, there is a 2×2 variance-covariance (VCV) matrix for each level of A. A chi-square test that Winer labels as chi-square 1 (Winer,

TABLE 3.2. Summary Table of ANOVA of Analysis 4 $[RS_{10} \text{ in } (A_3)]J_2$

Source	df	*MS*	*F*	*p*
Between subjects				
Treatments (*A*)	2	106.07	5.79	<.01
Error, *RS*(*A*)	27	18.32		
Within subjects				
Pre-post (*J*)	1	7437.07	4803.83	<.001
AJ interaction	2	80.07	51.72	<.001
Error, *RSJ*(*A*)	27	1.55		

MS_E to use on between-subject simple effects = 9.9333, Satterthwaite df = 31

1971, p. 595) tests the homogeneity of these three VCVs and yields a value of 3.58 with 6 degrees of freesom. Thus we retain the null hypothesis that these three sample VCVs may be random samples from the same population VCV matrix, or that the population VCVs are homogeneous. Now we will obtain a summary table of greater complexity, as shown in Table 3.2.

Note the significant interaction of Table 3.2. The *F* for the *AJ* interaction is exactly the same as that for the treatments effect when analyzing gain scores in analysis 2, $F = 51.72$. This is no coincidence, but will happen every time. One suspects that some editors who may complain about the use of gain scores do not know that the two-factor analysis is doing exactly the same test in a new disguise.

The significant *F* on the *AJ* interaction tells us that the mean gain differs from group to group. The two significant main effects are usually of little interest, given a strong significant interaction. The $F = 5.79$ on treatments tells that the subject means (averaged over *J*1 and *J*2) differ from group to group. This is of little interest, since the subjects should be the same on the pretest and there is no logical reason to average the pretest and posttest. Similarly, the significant *J* main effect ($F = 4803.83$) merely tells us that, averaged over the three *A* conditions, the pretest mean is significantly less than the posttest mean. This test is trivial, since inspection of the data shows there is no overlap between the two sets of scores. The maximum score on the pretest is 9. The minimum score on the posttest is 16. When the scores do not overlap, the means are significantly different by the interocular test (i.e., the data hit you between the eyes), and the formality of the ANOVA test is unnecessary. However, the experimenter may do it anyway for the thrill of obtaining an $F = 4803.83$.

After a significant AJ interaction, the usual follow-up tests would test the simple effects of cell means. You may do a WSD on the pretest means (which retains the equality of the three means) and another WSD on the posttest means. In this case, the proper error term to use is not found directly in the summary table of Table 3.2. The $MS_{RS(A)}$ = error term for MS_A is based upon a linear function of the subject sums (or subject means). Since the subject sums are summed over both levels of J, they have a greater variability than that found in either column alone (given a positive correlation). Everybody should remember that the variance of the sum is $S_1^2 + S_2^2 + 2S_{1,2}$, where the first two terms are variances and the last term is two times the covariance of the two columns. If you just average the six cell variances alone, you obtain $MS_{w\text{ cells}} = 9.333$. Note that this MS_E is smaller than $MS_{RS(A)}$ and larger than $MS_{RSJ(A)}$. This will always be the case when the covariances are positive.

The df to use with this $\text{MS}_{w\text{ cells}}$ must be found by a formula by Satterthwaite (1946). Both outputs are given by the computer program ANOVR (Games, 1975, 1981, Games et al. 1980). Using this term, we obtain the WSD for the pretest column and for the posttest column:

$$\text{WSD} = q_{.01,3,31}\sqrt{\frac{9.9333}{10}} = 4.45(.99667) = 4.435$$

Applying this to the pretest means again establishes that they are not significantly different. Applying this to the posttest means, we find that the $A3$ mean (22.5) is significantly less than the other two means, but the other two means are not significantly different. Recall that we had enough power in analysis 2 to show that the three gains were significantly different, but here we have results similar to the first analysis, where the $A2$ and $A3$ means did not differ significantly. This is because of the size of the error term. The MS_E of the gain analysis was 3.10, thus giving lots of power. The MS_E of analysis 1 on the posttest alone was 11.13, giving much less power. In the present analysis, $MS_{w\text{ cells}} = 9.933$ is similar to the first error term. The first error term averaged the three variances for the posttest column alone, giving 11.13. The present analysis averages the variances of all six cells, giving us 9.933. We would expect these two terms to be about the same, as they are here. For completeness, note that the between-subjects error term, $MS_{RS(A)} = 18.32$, about twice as large.

With positive covariances, we may always expect these three error terms to have the following relative sizes: $MS_{RSJ(A)} = MS_{Ew} < MS_{w\ \text{cells}} < MS_{RS(A)} = MS_{Eb}$. Hence, ignoring the small differences associated with differing df, (with positive covariances) we may always expect the greatest power on within-subjects factors (here J and AJ effects, and J simple effects), and medium power on between subject simple effects (here A effects at $J1$ or $J2$). The largest error term (with positive covariances) will be $MS_{RS(A)}$, which is used with the corresponding marginal or main A means. However, those means are based on 20 observations, while the simple effects using cell means are based on only 10 observations. Of course, the larger n tends to increase power while the large MS_E tends to decrease it.

The usual procedure after a significant interaction is to do tests on simple effects, using the cell means. We have already commented on the simple effects of A at $J1$, and at $J2$. The other way is to test the $J1$ mean versus the $J2$ mean for each level of A. This is equivalent to testing that the gain means are zero. All are significantly greater than zero ($p < .001$).

5. Another method of analysis is to use the pretests to define a fixed organismic (that is, nontreatment) factor. The experimenter should stratify on the pretest into upper and lower groups, then assign subjects randomly to the two treatments. This should be done before the treatments are administered. Actually, this was done here, note that the first five subjects in each group have pretest scores of 5–9, whereas the last five have pretest scores of 0–4. This type of design is often called a *treatment-by-levels design,* or *treatment by blocks.* Note that subjects cannot be randomly assigned to the organismic factor; all we can do is obtain the subject's score and then classify him or her as belonging to one block or another. We might represent this design by $R_L S_5$ in $(L_2 A_3)$, or perhaps $RS_5 wL$ in $(L_2 A_3)$. Either notation is intended to indicate that the random assignment of subjects was true for A, but not for L.

Now when the analysis is conducted with this additional block, we obtain the summary table in Table 3.3.

Note that $MS_A = 183.63$, exactly as in analysis 1 because we again are using only the posttest scores, and the three A means of the treatments have not changed. However, $MS_{w\ \text{cells}} = 6.67$ now, but it was 11.13 in analysis 1. This is because dividing the subjects into the upper and lower levels (based on the pretest) has reduced the posttest variance within the cells. This reduction in the error term increases the power of the F-test on the A treatments, and the power of the WSD. The WSD $= 4.54\sqrt{6.76/10} = 3.733$, down from 4.790 in analysis 1. This is an example

TABLE 3.3. Summary Table for Treatment(3) by Blocks(2) Analysis

Source	df	*MS*	*F*	*p*
Between subjects				
Treatment (*A*)	2	183.63	27.14	<.001
Blocks (*L*)	1	136.53	20.18	<.001
AL int.	2	0.83	0.12	
Within cells				
RS(*LA*)	24	6.67		

of using a blocking factor (levels) as a means of reducing the error term, and increasing power. In addition, we can now see if there is a levels by treatment interaction. In this case, the interaction terms has $F < 1$, so there is no suggestion of an interaction.

In this data set, the treatment-by-levels analysis does not reduce the error term as much as the gain score analysis, or the covariance. This is because of the high correlation between the pretest and posttest. The optimum number of levels to maximize power varies with the correlation. In this case, more than two levels would be appropriate. However, this author recommends never going below two subjects per cell in such an analysis, so that you always will have a test of the treatment-by-levels interaction, and always will have a within-cells term in your analysis. Myers (1979, pp. 150–155) has a section on the optimal number of levels, attending only to power considerations. You might go up to five levels with only two subjects per cell, in this case. However, such decisions should be made a priori, and since they were not, we shall not attempt to break up the levels on a post hoc basis. The main point is that the relative efficiency of the treatment by levels analysis, the covariance analysis, and the gain score analysis will vary with the size of the correlation between the pretest and posttest. for details, see the Myers reference.

6. It is possible to do a treatment-by-levels analysis of the gain scores, again using the pretest to create two levels. This is largely a waste of time, because the correlation of the pretest to the gains is low, and there is little reduction in MS_E. We shall not pursue this analysis any further. This merely illustrates there is no point in blocking (matching) if the blocking variable does not have a decent correlation with the criterion. The author has sometimes been dismayed by students who want to match on height and eye color, when there is no evidence to conclude that either of these organismic variables correlates with the criterion behavior. There is no intrinsic virtue in matching in the absence of such evidence.

It is possible to create even more complex analyses of this rather simple data set. For example, one student suggested using the treatments-by-levels analysis of part 5, but also using the pretest as a covariate. Since the covariant (pretest score) was used to create the levels, it is obvious that the means of the covariate will differ substantially between groups. Refer to Cochran, (1957), Elashoff (1969) and Games (1975) as to why this is a disastrous suggestion.

2. Three Experimental Treatments with Four Levels of a Repeated Measure

Let us go to a more complex data set with more subjects, and more levels of the repeated-measures factor. More subjects are desired in terms of stability of results, and more levels of *J* are desired in order to obtain MANOVA statistics that will not all be the same. Table 3.4 contains the raw data of an $[RS_{15}$ in $(A_3)]J_4$ design.

With more than two levels of the repeated-measures factor *J*, new

TABLE 3.4. Raw Data Set for ANOVA & MANOVA Comparisons $[RS_{15}$ in $(A_3)]J_4$

*A*1 Treatment				*A*2 Treatment				*A*3 Treatment			
*J*1 pre	*J*2 post	*J*3 ret 1	*J*4 ret. 2	*J*1 pre	*J*2 post	*J*3 ret. 1	*J*4 ret. 2	*J*1 pre	*J*2 post	*J*3 ret. 1	*J*4 ret. 2
54	57	58	67	48	56	53	51	56	56	56	57
52	54	64	68	52	60	67	63	48	48	47	49
47	48	58	64	50	57	58	59	47	45	44	46
52	52	61	68	56	69	69	65	60	61	62	59
47	48	55	61	47	59	56	55	54	57	61	59
52	53	58	68	53	62	61	59	60	59	59	55
56	57	62	66	49	59	69	53	47	44	43	47
51	55	64	64	51	60	64	59	49	48	46	47
47	49	59	70	49	56	56	59	47	47	48	47
51	57	65	71	49	61	60	65	49	51	51	50
48	48	52	60	51	60	58	61	49	48	49	45
48	49	55	66	52	61	59	59	51	52	52	57
49	52	59	64	58	69	66	65	47	43	43	43
51	52	59	66	47	55	51	53	53	50	56	53
57	61	67	73	48	54	56	53	51	46	46	49

assumptions come into play in the univariate analysis. See Myer pp. 162–178). With four levels of J, we now will have a 4×4 VCV matrix for each level of A. The assumption is made that the population VCVs are homogeneous, so that the sample VCVs differ only by sampling fluctuation. This may be tested by a multivariate chi-square test, Chi-Square 1. Equal n's are desirable to give robustness against violations of this assumption. The Chi-Square 1 test on the data of Table 3.4 yields a chi square of 19.416 with $\text{df} = 20$ and $p = .496$, so this null is retained. Retaining this null encourages one to use the pooled VCV matrix as the best estimate of the common VCV matrix. When you retain the null this is encouraging. Unfortunately, Rogan, Keselman, and Mendoza (1979) show that this test yields a surplus of Type I errors if the Y-distributions sampled are leptokurtic. However, if you retain the null, you are not bothered by an inflated risk of Type I errors. We used this test, and corresponding assumption in analysis 3, so this is not the new one.

The new requirement is that this pooled VCV matrix has a special form known as sphericity, so $S_j^2 + S_{j'}^2 - 2S_{jj'}$ is a constant. When this condition is met, a coefficient known as *epsilon* ε is equal to 1.0. This was first defined by Box (1954), and a sample estimate of the parameter is usually credited to Greenhouse and Geisser (1959). In general, $1/(K-1) \leq \varepsilon \leq 1.0$. The lower the index, the further away the VCV matrix is from the desired condition of sphericity. The term $1/(K-1)$ is the lower limit of ε, and 1.0 is the upper limit (K is the number of levels of the repeated measure factor J). Box showed that MS_J/MS_{RSJ} is distributed as F with $\text{df}_u = \varepsilon(K-1)$, and $\text{df}_d = \varepsilon(K-1)(n-1)$ in a design with J as the repeated measure factor and n random subjects (or random blocks). Here df_u = degrees of freedom upstairs and df_d = degrees of freedom downstairs. Greenhouse and Geisser (1959) extended this to the present design. Essentially, in the univariate analysis, you should adjust df_u and df_d of the statistic by multiplying the conventional df by the sample estimate of ε. Collier et al. (1967) showed that this process yields an approximately unbiased test.

Unfortunately, this is one of the great neglected areas of statistical applications. Dozens of computer programs for doing such analyses completely neglect this problem, and report the probabilities of MS ratios using the F-distributions with conventional df's (implicitly assuming that $\varepsilon = 1.0$). Using the conventional df's results in seriously positively biased tests that have a great surplus of Type I errors. Naturally, most people trained in the 1950s did not know of the ε-factor, and many journals have published thousands of articles with these seriously flawed tests, resulting in hordes of Type I errors in the literature. Even outstanding statistical packages like BMDP and SAS have added the ε index and the adjusted df to their programs only in the 1980s.

Using the ANOVR program, and the data of Table 3.4, we get the pooled VCV matrix and summary table of Table 3.5.

TABLE 3.5. Means, Pooled VCV Matrix and Summary Table of Table 3.4

Means	$J1$	$J2$	$J3$	$J4$
$A1$	50.80a	52.80b	59.73c	66.40
$A2$	50.67a	59.87	60.20c	58.60
$A3$	51.20a	50.33b	50.87	50.87

Pooled VCV matrix and related indices

	1	2	3	4
1	13.5270			
2	15.8984	22.2730		
3	15.7287	21.9825	30.5016	
4	12.3952	17.2873	18.9651	21.1651

Mean of the diagonal entries (variances) = 21.8667
Mean of the off-diagonals (covariances) = 17.0249
Mean of the variances − mean of the cov. = 4.8238 = $MS_{RSJ(A)}$
Mean of the variances + $(K-1)$ *mean of cov. = 72.9952 = $MS_{RS(A)}$
Epsilon hat = Greenhouse & Geisser statistic = .8179

Above VCV matrix converted to a correlation matrix

	1	2	3	4
1	1.000			
2	.916	1.000		
3	.774	.843	1.000	
4	.733	.796	.746	1.000

Summary table

Source	df	*MS*	*F*	conv. *p*	adj. df *p*
Between subjects					
A	2	862.5722	11.817	<.001	
$RS(A) = Eb$	42	72.9952			
Within subjects					
J	3	510.7907	105.889	<.001	<.001
AJ interact.	6	275.1796	57.046	<.001	<.001
$RSJ(A) = Ew$	126	4.8238			

MS_E for simple effects of A at levels of $J = 21.8667$ with Satterthwaite df = 59.

Note the shape of the VCV matrix and correlation matrix. This is typical of learning data, where the greater the number of trials (or time) between two columns, the lower the correlation (Fleishman, 1960; Games, 1962). There are even theories about such learning matrices based upon such data, since they show this pattern so regularly (Humphreys, 1960; Jones, 1962). The data of Table 3.4 are not a good example to show the increased risk of Type I error when using conventional probabilities, since the present *MS* ratios are so large, they are significant even at the conservative adjustment by using $\varepsilon = 1/(K-1)$.

Note the relationship of the MS_E's in the summary table to the variance covariance matrix. $MS_{RSJ(A)} = MS_{Ew}$ often referred to as $MS_{\text{error within}}$ is equal to the mean of the variances *minus* the mean of the covariances. The $MS_{w\text{ cells}}$ equals the mean of the cell variances, 21.86667, and is the appropriate MS_E if one is testing the simple effects of A at a given level of J. Finally, the mean of the variances *plus* $(K-1)$ times the mean of the covariances is equal to $MS_{RS(A)} = MS_{Eb}$, or the error term for the between-subjects portion of the analysis. This author's experience has been that, on behavioral data, he almost always has encountered positive covariances (unless the experimenter has been foolish enough to use scores that are ipsative over J). With positive covariances, it automatically follows that

$$MS_{RSJ(A)} = MS_{Ew} < MS_{w\text{ cells}} < MS_{RS(A)} = MS_{Eb}$$

so that we usually will have the greatest power on the within subjects tests.

Since we have a significant interaction, conventional logic suggests we look at the cell means as simple effects. Again the Tukey WSD is a convenient way to do so. As before, we shall use the .01 level of significance. For the J effects at a given level of A, we need

$$\text{WSD}_{J|A} = q_{.01,4,126}\sqrt{\frac{MS_{Ew}}{15}} = 4.50\sqrt{\frac{4.82381}{15}} = 2.552.$$

Applying this result to the means on Table 3.5 yields the underlinings shown. We see that $A3$ (control group) has no significant changes. However, $A2$ jumped to a higher value on the posttest ($J2$) and remained at this level on both retention tests ($J3$ and $J4$). However, the $A1$ group made only a slight nonsignificant gain on the posttest, but continued rising on both retention tests, reaching the highest mean obtained in the experiment at $J4$.

The WSD for the A columns at given J's is found by the following:

$$\text{WSD}_{A|J} = q_{.01,3,59}\sqrt{\frac{MS_{w\text{ cells}}}{15}} = 4.37(1.20738) = 5.276.$$

Applying this to the columns of means in Table 3.5, we find no significant differences in the pretest (as expected), so we will place an "a" to the right of each $J1$ mean, indicating that those means that share the same letter are not significantly different. This is an alternative way to do the same thing that Duncan's underlining notation does, indicating those means where we retain the null. In the $J2$ column, the $A1$ and $A3$ means are not significantly different, so we join them by placing b's to the right. In the $J3$ column, the $A1$ and $A2$ means are not significantly different, so they receive c's to the right. Finally, in the $J4$ column, all means are significantly different, so there are no letters to the right of any means. Notice how concisely the experimental results are summarized in this notation. If you understand the notation, looking at the modified table of means conveys the major results.

The preceding use of the WSDs has one major flaw. It assumes all the assumptions have been met. It is possible to use the obtained VCV matrix instead of a single composite error term (Mitzel and Games, 1981; Maxwell, 1980) either when doing pairwise contrasts (as in the WSD) or with complex contrasts. However, we shall not get into the details here, but will instead consider the multivariate tests that may be used instead of the univariate test illustrated in Table 3.5.

SAS PROC GLM is a general linear model program that can do the foregoing analysis. In the present experiment, with 45 subjects nested in three levels of A with each subject yielding four observations, we have 180 values of Y. Obviously, these 180 observations must be "punched" (or entered into the computer; saying "punched" is briefer). In the past, before SAS version 5 (SAS V5) came out in 1985, it was necessary for the user to input each Y observation separately with each level of each factor overtly identified. Thus for each Y, we also had to punch a subject identifier, an A identifier, and a J identifier, or each Y actually required four punched numbers. Thus 720 different values would have been punched in this case. Fortunately, SAS V5 contained a new way to handle repeated-measures factors that greatly reduced the amount of punching required. Using one card per subject (or one row per subject) is a natural way to get data into the computer, and one acceptable to most computer packages. It now is acceptable for SAS PROC GLM when you have a repeated-measure design. You still need a subject identifier, and an *A* identifier, but the four *Y*-values for each subject may be placed on the same row. Thus you will need only 45 subject identifiers, 45 *A* identifiers, and your 180 *Y*-values for a total of 270 punched values, greatly reducing the pain of input. (It is possible to reduce the amount of punching further by leaving some factors as implicit rather than overt, but it is much safer to do the preceding.)

Let us assume you have read in the data using *RS* as the label of the column of subject identifiers, using *A* as the label of the column of treatment identifiers (1, 2, or 3). Label the pretest (*J*1) column as *Y*1, the posttest (*J*2) as *Y*2, the first retention test (*J*3) as *Y*3, and the second retention test (*J*4) as *Y*4. Your minimal SAS instructions to run this job are

```
PROC GLM; CLASS A;
MODEL Y1 Y2 Y3 Y4 = A; REPEATED J 4/ SHORT;
LSMEANS A /STDERR PDIFF; MEANS A /TUKEY ALPHA = .01;
```

If you have never run SAS before, the first rule is that every command must end with a simicolon. When you leave off semicolons, you often get weird error messages, so the first thing to check when you get error messages is 'were all the semicolons there?"

PROC GLM just loads this particular procedure into the computer. I'm assuming you just read in this data set, so it will automatically be used. If you have several data sets, you may specify which one you want to use. CLASS A; identifies the *A* variable as a factor you will use in the ANOVA. MODEL Y1 Y2 Y3 Y4 = A ;REPEATED J 4 tells the computer that you are using *A* as a factor, and the four *Y*'s are really four levels of the repeated-measure factor *J*. Thus these two statements tell the program you have two crossed factors, *A* and *J*, so that the *AJ* interaction will automatically be done. You may ask where the *RS* factor went. It is here left implicit, since if you make it explicit, you will obtain $MS_{\text{error}} = 0$, which will delete all the *F*'s (and associated probabilities), since you cannot divide by 0. All multivariate tests would be deleted also.

We did not complete the repeated statement. The entire line is REPEATED J 4 / SHORT; The / indicates that options to the REPEATED command follow. The option used asked for the SHORT version of the output of the multivariate tests. If you have never used SAS PROC GLM before, you might skip this on the first job you run to get as complete an explanation as possible. After the first job you probably will want to use it, unless you like lugging tons of paper. Even with the SHORT output, SAS prints nine pages of output from this small instruction set. Without the SHORT instruction, it prints twelve pages of output. SAS permits you to delete and/or include many different outputs, but we shall concentrate on the "default output," that which SAS automatically gives with minimal options specified.

The first output is a univariate analysis of *Y*1 as an RS_{15} in (A_3) design. This is an analysis of the simple effect of *A* on *J*1, like the WSD done before after the significant *AJ* interaction. However, this is done by an *F*-test instead of the WSD test. Both the *F* and the WSD control the familywise level of risk of Type I error (Games, 1971) at the level of significance ($\alpha = .01$ in this chapter). However, they use a somewhat different region of retention (Games, 1971), so they may differ in outcome on borderline cases.

This univariate analysis is then repeated for *Y*2, *Y*3, and *Y*4. In the previous WSD we used a common error term throughout. Here, new error terms will be computed for each separate analysis, so this also may make results differ from the WSD. In this example, the *F* on the pretest ($Y1 = J1$) is nonsignificant, but the other three are significant with *F*'s of 13.6 to 42.75. The $MS_{RS(A)}$ values for the four *Y*'s are 13.527, 22.273, 30.301, and 21.165. You may recall those as the four variances found on the diagonal of the VCV matrix of Table 3.5. These were then used in the four *F*'s before. Recall that the average of these four variances is the $MS_{w\ \text{cells}} = 21.8667$ we used to do the WSDs after the analysis of Table 3.5.

After the four univariate ANOVAs, with the MEANS statement given earlier, you will obtain four outputs labeled "Tukey's studentized range (HSD) test for variable *Y*1, and a repeat for variables *Y*2, *Y*3, and *Y*4." The HSD label was indeed Tukey's, standing for "honestly significant difference." This strikes the author as pretentious, since it would imply that all the other solutions to the multiple comparisons problem are "dishonest." Thus we will use Miller's (1981) label of WSD for "wholly significant difference." These HSD results from SAS are like the WSD results we obtained except for the fact that they also are based on four different error terms. They also are given vertically using letters to indicate retentions. For *Y*2, the following is printed:

```
NOTE: THIS TEST CONTROLS THE TYPE I EXPERIMENTWISE
  ERROR RATE BUT GENERALLY HAS A HIGHER TYPE II
  ERROR RATE THAN REGWQ
    ALPHA = .01 DF = 42 MSE = 22.273
```

CRITICAL VALUE OF STUDENTIZED RANGE = 4.37
MINIMUM SIGNIFICANT DIFFERENCE = 5.325
MEANS WITH THE SAME LETTER ARE NOT SIGNIFICANTLY DIFFERENT

TUKEY GROUPING	MEAN	N	A
A	59.867	15	2
B	52.800	15	1
B	50.333	15	3

This tells us that the $A2$ mean is significantly higher than either the $A1$ mean or the $A3$ mean, as we saw before. The author prefers referring to the FAMILYWISE risk of Type I error rather than the EXPERIMENTWISE in that this set of three means by no means constitutes the whole of our statistical inference in this entire 3×4 experiment.

This is repeated with different error terms for all the Y's. In general, it is safer when using PROC GLM to ask for LSMEANS rather than MEANS. The reason is that, when you have unequal n's, LSMEANS give you "predicted means" that correspond to the Type III sum of squares and mean squares you should use in your summary table. For unequal n cases, the results of a MEANS command may correspond not at all to your summary table. Also when doing an analysis of covariance, LSMEANS must be used to give you the adjusted means. However, for equal n cases of ANOVA, as here, LSMEANS and MEANS give identical values of the means, and the MEANS statement has far more choices of multiple comparison options.

After the MEANS output, the LSMEANS output is given. We will cover this type output when discussing covariance. Here it agrees with the MEANS output, except it does not do a WSD.

Next is the important multivariate test of the J main effect means (testing the equality of the marginal J means). If you did not use the /SHORT option; SAS prints

MANOVA TEST CRITERIA FOR THE HYPOTHESIS OF NO J EFFECT

followed by a whole page that gives you many different values related to four different multivariate statistics testing this null hypothesis. If you did use the /SHORT option, you will receive a longer title, but a shorter output of only 11 lines. This will read as follows:

MANOVA TEST CRITERIA AND EXACT F STATISTICS FOR THE HYPOTHESIS OF NO J EFFECT
H = TYPE III SS&CP MATRIX FOR:J
E = ERROR SS&CP MATRIX S = 1 M = 0 N = 20

STATISTIC	VALUE	F	NUM DF	DEN DF	PR > F
WILKS' LAMBDA	.2800382	52.704	2	41	.0001
PILLAI's TRACE	.7199618	52.704	2	41	.0001
HOTELLING-LAWLEY TRACE	2.570941	52.704	2	41	.0001
ROY'S GREATEST ROOT	2.570941	52.704	2	41	.0001

If you are a novice to multivariate analyses, you may ask why SAS bothers giving you four different statistics if all reduce to exactly the same F-value. The answer is that it is only for this test of the main effects of the J-factor that all four will give you exactly the same value of F. One of the virtues of the longer output is that it will refer you to some references in the SAS GLM chapter that take you back to the original literature. Each of these statistics (as in the first column under VALUE) can be referred to other tables. In later tests, only the Wilks' lambda will convert to an exact F-value; the others will convert to an approximate F-statistic.

You may recall that the univariate F for the main effect of J yielded $F = 105.89$. Now we get $F = 52.704$, just about half the size. Obviously, both are significant, but with more borderline data sets with smaller n's, why would you want to use the multivariate F?

The biggest reason is that the multivariate F does not have to assume anything about the Box ε index. You recall we recommended computing $\hat{\varepsilon}$ and multiplying the original df by $\hat{\varepsilon}$ to obtain an approximate F-test that is relatively unbiased. The present multivariate F is an exact test (under the assumption of multivariate normality) that does not have to deal with the ε index. Unfortunately, we do not know how robust the multivariate test is to violations of multivariate normality, nor do we have good ways of detecting multivariate *non*normality. It may be better to regard the multivariate tests as approximate tests on real data, also.

Actually, the four different multivariate tests will differ in both power and robustness. In general, there are several reasons to avoid Roy's greatest root test, as it will have a lack of power in many situations. Each of the other three tests might have maximal power in various situations, and lower power in other situations. Mardia (1971) and Olson (1974) show that Pillai's trace test may be most robust against various forms of multivariate *non*normality. A general reference on the power of the three tests is Schatzoff (1966). Unfortunately, power results are very complex, and usually the experimenter does *not* know the parameter condition so well that he or she can tell which test would be most powerful.

We would also be interested in comparing the power of the univariate test using $\hat{\varepsilon}$ with the multivariate test. Unfortunately, the only study the author knows of compared the multivariate test with the most conservative version of the univariate test, merely using $\hat{\varepsilon}$ as $1/(K-1)$ (Davidson, 1972), so we have no good data on which to make a power comparison. Even with better data, the power relation may vary with conditions, so it may be hard to make a choice.

Let us continue to the multivariate test of the *JA* interaction. We shall skip the usual heading, which is the same as the previous one except for *JA*.

S = 2 M = 0 N = 19.5

STATISTIC	VALUE	F	NUM DF	DEN DF	PR > F
WILK'S LAMBDA	.02737402	67.254	6	80	.0001
PILLAI'S TRACE	1.666812	68.369	6	82	.0001
HOTELLING-LAWLEY TRACE	10.17169	66.116	6	78	.0001
ROY'S GREATEST ROOT	5.79753	79.233	6	41	.0001

NOTE: F STATISTIC FOR ROY'S GREATEST ROOT IS AN UPPER BOUND. F STATISTIC FOR WILKS' LAMBDA IS EXACT

Now we have reached the more typical case where the multivariate tests no longer yield exactly the same *F*'s. The messages at the bottom give us some guidance. Roy's greatest root yields an upper bound on *F*, so that this test is positively biased; it will yield more Type I errors than alpha. In the present data set, the effects are massive, and all four tests would lead us to reject the null of no *JA* interaction. However, in borderline cases, Roy's test would be a poor choice. The other three tests do not have an automatic preference. All will control the risk of Type I error around α, given multivariate normality, and which is most powerful varies with factors beyond the experiments control. There might be a slight preference for Wilks' test, because it converts to an exact *F*-statistic, while the other two are only approximations. Unfortunately, on borderline data, one suspects that many experimenters choose the alternative that yields the most significant results, either by the largest *F* or the smallest post hoc probability.

These are the two within-subjects factors in this repeated-measures design, and these are the two multivariate tests for testing them. The foregoing output then corresponds to the within-subjects portions of the summary table from ANOVR in Table 3.5. The next SAS output should exactly match the "between-subjects" portions of Table 3.5. The output is

TESTS OF HYPOTHESES FOR BETWEEN SUBJECTS EFFECTS

SOURCE	DF	TYPE III SS	MEAN SQUARE	F VALUE	PR > F
A	2	1725.144444	862.5722222	11.82	.0001
ERROR	42	3065.800000	72.9952381		

This should closely agree with the conventional ANOVA from ANOVR or any other program. Some difference in right-hand digits might result from differing computer accuracy in different algorithms. Now SAS gives you the univariate repeated-measures analysis. The output is as follows:

						ADJUSTED PR > F	
SOURCE	DF	TYPE III SS	MEAN SQUARE	F VALUE	PR < F	G–G	H–F
J	3	1532.372222	510.79074	105.89	.0001	.0001	.0001
JA	6	1651.077777	275.17963	57.05	.0001	.0001	.0001
ERROR(J)	126	607.800000	4.82381				

GREENHOUSE–GEISSER EPSILON = .8174,
HUYN–FELDT EPSILON = .9138

So we see that SAS PROC GLM will give you the univariate tests on the repeated measure factor and its interaction, the same as ANOVR; it just will not give you the VCV matrix or correlations between the repeated measures. The Greenhouse-Geisser (1959) ε should match the ANOVR epsilon. The Huynh–Feldt (H–F)ε is a slightly different statistic. The Greenhouse–Geisser (G–G)ε is the maximum likelihood estimate of Box's parameter ε. Huynh and Feldt (1970) and Rouanet and Lepine (1970) showed that the G–G ε is somewhat conservative at small n's. Unfortunately, the H–F epsilon is somewhat liberally biased for small n, yielding more type I errors than expected. The author prefers the G–G. SAS computes three probabilities for this case, the first with the classical df (seriously positively biased), the second adjusting the df with the G–G, and the last adjusting the df with the H–F. The author would suggest always using the G–G probability. In this case, it makes no difference, but on small n cases with low power, the difference is often not trivial. The H–F index sometimes exceeds 1.0. In this case, it must be rounded down to 1.0, since this is the upper limit of the parameter.

It is possible to obtain the VCV matrix from SAS using PROC DISCRIM or PROC STEPDISC. This is very sensible since discriminant analysis is a logical inverse of the MANOVA. In MANOVA you have groups defined by treatments, and ask if the several variables have differing means. In discriminant analysis you have several variables, and ask if you can use linear combinations of these several variables to separate the groups. In general, you should have a significant MANOVA before bothering to do a discriminate analysis for the purpose of assigning subjects to groups.

In PROC DISCRIM you should use several options, as follows:

```
PROC DISCRIM SIMPLE WCOV WCORR PCOV PCORR POOL = TEST;
```

SIMPLE will give you the descriptive mean and variance on each variable for each group. WCOV will print the VCV matrix for each group (there is an option in ANOVR that will do this also). WCORR will print the within group correlations. PCOV will print the pooled VCV matrix (like the one automatically done by ANOVR). The PCORR command converts the pooled VCV matrix into a correlation matrix. The print labels this correlation matrix as a "partial" correlation matrix. That particular word *partial* is rather dubious. It may be defended in that the *A*-factor has not been allowed to influence the correlations, so they are partial *r*'s with respect to the *A*-factor. They are not partial *r's* like some where $r_{12.34}$ would stand for the partial correlation of *Y*1 and *Y*2, with *Y*3 and *Y*4 partialed out. POOL = TEST will produce the chi-square 1 test of ANOVR, with considerably more detail. Of course, there is considerably more output given in that every subject is assigned to a group, and some are properly classified, while others are misclassified. In a real experiment, subjects are randomly assigned to treatments, and there is no interest in reversing the process. Discriminate analysis is appropriate where you have several organismic variables and you wish to use these to classify subjects in diagnostic categories. For example, patients in a mental hospital might be classified into categories of manic-depressive psychosis, schizoprenia psychosis, paranoid schizophrenia psychosis, and so on. Hopefully, if dfferent treatments are successful with these different classes of patients, then it would be useful to classify entering patients using discriminant analysis. This is getting us remote from ANOVA and MANOVA, so we shall drop the topic. Recall the point was that SAS PROC GLM will not give you the VCV matrices, but PROC DISCRIM will. Unfortunately, the author's experience suggests

TABLE 3.6. Outputs on the Gain Score Analysis of, Say, 2

Means = LSMEANS

	$G2$	$G3$	$G4$	$\bar{Y}_{j.}$
$A1$	3.000	5.933	4.600	4.511
$A2$	2.200	6.533	5.933	4.889
$A3$	3.133	4.667	3.667	3.822
	2.777	5.711	4.733	$4.407 = \bar{Y}_{..}$

MANOVA for the hypothesis of no J effect

Wilks' lambda = .44218 $F(2, 41) = 25.861$ $p < .0001$

MANOVA for the hypothesis of no JA effect

Wilks' lambda = .76409 $F(4, 82) = 2.952$ $p < .0248$

Tests for hypotheses for between subjects effects

Source	df	Type III SS	Mean square	F value	$PR > F$
A	2	26.32592	13.16296	0.82	.4463
$E = RS(A)$	42	672.26667	16.00635		

Univariate tests of hypotheses for within subject effects

Source	df	Type III SS	Mean square	F value	$PR > F$
J	2	200.77037	100.38518	19.18	.0001
JA	4	47.49629	11.87407	2.27	.0686
$E = RSJ(A)$	84	439.73333	5.23492		

G–G epsilon = .9113; H–F epsilon = .9958

that very few SAS users of ANOVA or MANOVA bother to obtain the VCV matrices.

One minor point: SAS PROC STEPDISC uses Wilks' lambda as the criterion that is maximized at each step (using the forward or stepwise options). Thus, for consistency, we might choose to use Wilks' lambda as the usual criterion in MANOVA in the absence of any clear reason for selecting one of the other two relatively unbiased criteria. For brevity, we shall report only the Wilks' lambda results in MANOVA's from Table 3.6 on. This is not meant to imply that the Wilks' test would always be preferred over Pillai's trace test or Hotelling's trace test.

To return to our MANOVA example of the data in Table 3.4, in case you did not notice, the multivariate F's for the AJ interaction were in the 60's, while the univariate F for the interaction was 57.05. Alas, it is not always possible to predict which will be greater.

There is one major danger using SAS PROC GLM on a repeated-measures design. If you use the REPEATED command, it will yield a multivariate analysis on the within-subject portions of the analysis as documented before. *Never* get carried away and insert a MANOVA statement for the between-subjects factors. If you insert the statement MANOVA H = A; in this design, you will get a very misleading output. The output will look reasonable, but it is *not* testing the hypothesis of no difference between the *A* marginal means (the *A* main means). The output looks lke this:

MANOVA TEST CRITERIA AND EXACT F STATISTICS FOR THE HYPOTHESIS OF NO OVERALL A EFFECT
H = TYPE III SS&CP MATRIX FOR:A
E = ERROR SS&CP MATRIX
S = 2 M = 0.5 N = 19

STATISTIC	VALUE	F	NUM DF	DEN DF	PR > F
WILKS' LAMBDA	.02346077	53.91	8	78	.0001
PILLAI'S TRACE	1.68873814	54.25	8	80	.0001
HOTELLING-LAWLEY TRACE	11.2673322	53.52	8	76	.0001
ROY'S GREATEST ROOT	6.80890365	68.09	4	40	.0001

Unfortunately, this test does *not* correspond to the test we want on the marginal means of *A*. Instead, it is testing if there are any significant differences between the three *A* means in any of the four vectors of *Y*1, *Y*2, *Y*3, and *Y*4. You could have a case where the *Y*1 score had *A* means of $A1 = 40$, $A2 = 50$, $A3 = 60$, while the *Y*2 means were $A1 = 60$, $A2 = 50$, $A3 = 40$. Assume no *A* differences in *Y*3 and *Y*4. Averaging these means over the four *Y*'s would give exactly 50 for each of the *A* marginal means. Since all of the *A* marginal means are exactly alike, we should obtain an *F* of 0. This is exactly what the univariate test of *A* would give. However, the multivariate test of *A* would give a large *F* on *A*, because the multivariate test is asking if there are any nonchance *A* differences in any of the four columns. Of course, there are such differences in *Y*1 and in *Y*2. Since the trends in the means are in the opposite direction, in the univariate analysis (and the REPEATED analysis) this would show up in the *AJ* interaction, where it belongs.

If this is confusing and you do not understand this last paragraph, please take my word for it. Any time you are using the REPEATED statement in SAS PROC GLM when you have only one criterion variable, do *not* use the MANOVA statement on any between-subject factors. Disaster awaits if you do.

Now we will do the gain score analysis. We define the three gain scores by subtracting the pretest score from each. In your SAS data step you should have three statements:

$$G2 = Y2 - Y1; \qquad G3 = Y3 - Y1; \qquad G4 = Y4 - Y1;$$

Since the $Y1$ means are about 50, and the $Y2$–$Y4$ means vary from 50.8 to 66.4, obviously the gain score means are going to be much smaller than the means reported in Table 3.5. We must change our SAS commands to reflect that we are using the three gain scores instead of four Y's. The SAS commands follow:

```
PROC GLM; CLASS A RS;
MODEL G2 G3 G4 = A; REPEATED J 3 /SHORT;
LSMEANS A/STDERR PDIFF;
```

This will produce the three univariate analyses on $G2$, $G3$, and $G4$. Then come the cell means as a 3×3 matrix, with A for rows and G for columns. These are given in Table 3.6, with the marginal A means and the marginal J means computed by hand.

You may recall that in our first example data, the main effect of A computed on the gain scores was exactly equal to the AJ interaction in the complete ANOVA of the [RS in ($A3$)]$J2$ design. This is true only for the case of two levels of the repeated-measure factor, since then the df of the A gain score analysis is 2 and the df of the AJ interaction is 2. In our more complex second example, the AJ interaction has df = 4, so we cannot expect exact correspondence as in Example 1.

In fact, so much has changed that it may not even be worthwhile spending time to relate Table 3.6 to Table 3.5. In Table 3.5 we had four levels of J; in Table 3.6 we have only three levels. In Table 3.6, we see that the difference between the marginal J means is significant by either the multivariate test, or the univariate test, using the .01 level of significance. The JA interaction would have been significant by the MANOVA test if we were using $\alpha = .05$, while we would have retained the null by the univariate test. With an $\alpha = .01$, we retain either way. The A main effect has $F < 1$, so retaining this null is clear. Conventional logic would suggest we merely do a WSD on the J marginal means to complete the analysis:

$$\text{WSD} = q_{.01,3,84}\sqrt{\frac{5.23492}{45}} = (4.28)(.34107) = 1.4598.$$

All the pairs of J marginal means differ by more than 1.4598 except G3 and G4, thus we underline these two means.

TABLE 3.7 ANOVA on $[RS_{15}$ in $(A_3)]J_3$ Design Using Posttest, Retention Test 1, and Retention Test 2

Means = LSMEANS

	$Y2$	$Y3$	$Y4$	$\bar{Y}_{J.}$
$A1$	52.600a	59.733b	66.400	59.644
$A2$	59.866	60.200b	58.600	59.555
$A3$	50.333a	50.866	50.866	50.688
$\bar{Y}_{.k}$	54.333	56.933	58.622	56.6296

MANOVA for the hypothesis of no J effect

Wilks' lambda = .22800382 $F(2, 41) = 52.704$ $p < .0001$

MANOVA for the hypothesis of no JA effect

Wilks' lambda = .1496998 $F(4, 82) = 32.484$ $p < .0001$

Tests for hypotheses for between subjects effects

Source	df	Type III SS	Mean square	F value	$PR > F$
A	2	2362.41481	1191.20740	18.77	<.0001
$E = RS(A)$	42	2665.73333	63.46984		
Univariate tests of hypotheses for within subject effects					
J	2	420.10370	210.05815	40.13	<.0001
JA	4	991.49629	247.87407	47.35	<.0001
$e = RSJ(A)$	84	439.73333	5.23492		

G–G epsilon = .9113, H–F epsilon = .9958

In order to provide a more meaningful comparison to Table 3.6, Table 3.7 contains the ANOVA on the $Y2$, $Y3$, and $Y4$ measures themselves. Of course, the means are much higher because we are not subtracting the pretest; we are merely ignoring the pretest. You may note that $MS_{RSJ(A)} = 5.2342$ in Tables 3.6 and 3.7. The reason is that when subtracting the pretest for each subject, you are subtracting a constant for the $Y2$, $Y3$, and $Y4$ of that subject. Subtracting a constant will not change $MS_{RSJ(A)}$ because we are using each subject as his or her own control in both analyses on the within-subject portions. Subtracting the subject pretest score has changed the cell means and marginal means, and, hence, the other MS's of Table 3.6.

The other term of interest is $MS_{RS(A)} = MS_{Eb}$. Note that on the raw scores (Table 3.7), $MS_{Eb} = 63.46984$. However, on the gain scores (Table 3.6) $MS_{Eb} = 16.00635$, which is roughly only 25% of the size of the same MS in Table 3.7. Just as in the first example, this reduction is due to a high correlation of the prescores with the posttest, retention 1, and retention two scores. It has thus reduced the error variance for A when working on the gain scores of Table 3.6. However, despite a smaller error term in Table 3.6, we see the A main effect is significant in Table 3.7 but not in Table 3.6. Looking at the main means of A shows that on the raw scores there are big differences between $A1$ and $A2$ versus $A3$. $A1$ and $A2$ have means of about 59, while the $A3$ mean is 50.68. However, on the gains, we see the A means do not differ by much, the smallest being 3.822 and the largest being 4.889. In general, we have very different results in Table 3.6 and Table 3.7. In Table 3.6, the JA interaction is nonsignificant, while in Table 3.7 it is significant.

Thus in Table 3.7, conventional logic suggests we do a WSD on the three cell means in a given row and on the three cell means in a column.

$$\text{WSD}_{J|A} = q_{.01,3,84}\sqrt{\frac{5.23492}{15}} = 4.28(.59076) = 2.5284.$$

While the means in the $A1$ row are significantly different, the $A2$ and $A3$ rows are underlined, indicating we retain the hypothesis of equal means.

To solve for the $MS_{w\ \text{cells}}$ for the $WSD_{A|J}$, we must pool the two error terms.

$$MS_{w\ \text{cells}} = \frac{SS_{Eb} + SS_{Ew}}{\text{df}_{Eb} + \text{df}_{Ew}} = \frac{2665.7333 + 439.7333}{42 + 84} = 24.64656.$$

Then

$$\text{WSD}_{A|J} = q_{.01,3,60}\sqrt{\frac{24.64656}{15}} = 4.28(1.28184) = 5.486.$$

Applying this in the $Y2$ column, we see the $A1$ and $A3$ means are not significantly different. Hence we follow the numbers with the common letter a. In the $Y3$ column, the $A1$ and $A2$ means do not differ significantly, so we follow them with the letter b. In the $Y4$ column, all the means are significantly different, so no marking is needed.

The remaining analyses of interest are the covariances on the Table 3.4 data. One of the first steps in any covariance is to determine if the assumption of homogeneous slopes is appropriate. If it is not, this indicates that different slopes are needed for each group. This is like an interaction; you would need a scatterdiagram with three different slopes, one for each group. Ignoring heterogeneous slopes is like ignoring a significant interaction. The easiest way to test for homogeneous slopes is to use the following SAS commands:

```
PROC GLM; CLASS A RS;
MODEL Y2 Y3 Y4 = Y1 A A * Y1 /SOLUTION ; REPEATED J 3 /SHORT;
```

This model emphasizes that the homogeneous slopes test is like an interaction test in that the A * Y1 symbolism is the same as for an interaction. You will again get all the univariate statistics on $Y2$, $Y3$, and $Y4$, but we are interested in the A * Y1 effect, which tests the hypothesis of homogeneous slopes. If we go to the univariate test, we get an $F_{2,39} = 0.84$ with $p = .441$, so we retain the null hypothesis of homogeneous slopes.

Both the univariate and the MANOVA outputs also give a test that many might not be expecting. Since PROC GLM treats the covariate ($Y1$) as another factor, when we include A*Y1 in the model we also get a three-factor interaction output, J*A*Y1. This is asking if the A*Y1 interaction changes from level to level of the J-factor, namely $Y2$, $Y3$, and $Y4$. Normally, we would have no reason to expect such a three-factor interaction, so it is comforting to see that we retain this null by either the MANOVA or univariate test. The MANOVA gives a Wilks' lambda = .9176854, which converts to $F_{4,76} = .834$, $p = .508$. The univariate test of this three-factor interaction also yields a nonsignificant F. Here the $F_{4,78} = 0.62$, $p = .651$. Thus we may breathe a sigh of relief that this crucial assumption is at least reasonable on the present data. As usual, retaining the null is not a strong proof, but if either of the foregoing tests were significant, preceeding with a conventional covariance would be very dubious logic.

Given that the assumption of homogeneous slopes is reasonable, we may go to the typical covariance model

```
PROC GLM; CLASS A RS;
Model Y2 Y3 Y4 = Y1 A/SOLUTION ; REPEATED J 3 /SHORT;
```

Note that it is helpful here to have the covariate term $Y1$ preceding the group term A. You must always keep in mind that PROC GLM (and its equivalents in other systems) proceeds in a hierarchical fashion from left to right. If you use a model statement with A $Y1$ than in a Type I SS, you will get the same SS_A that you get from an ANOVA without $Y1$ in the statement at all. A Type III SS_A will be OK because all other terms have been partialed out. In the preceding analysis, the model had $Y1$ A Y1*A and the term we were really interested in came last. It did not really matter if we had the $Y1$ first or second, as long as it preceded Y1*A. However, in a covariance, it is helpful to keep the $Y1$ first, so that the SS_A will be the SS for the adjusted A means, either as Type I SS or as Type III SS. Thus it probably is good practice to always keep the covariate as the first term in the model.

Now we will receive a set of univariate analyses, one on $Y2$, one on $Y3$, and one on $Y4$. Following this, we obtain a set of LSMEANS for each variable; then the MANOVA results, and finally the between-subjects and within-subjects univariate effects. These are reported in Table 3.8.

One feature of Table 3.8 is very shocking. Why has the test of no J effect gone down so drastically? The covariate is expected to influence only the between-subjects effects. Effects on the within-subject terms

TABLE 3.8. Adjusted Cell Means, MANOVA Results, and Summary Table for the Covariance of Table 3.4 Data, Using $Y1$ as the Covariate

Adjusted cell means

	$Y2=J2$	$Y3=J3$	$Y4=J4$
$A1$	52.9045	59.8367	66.4815
$A2$	60.1278	60.4584	58.8036
$A3$	49.9677	50.5049	50.5186

MANOVA for the hypothesis of no J effect

Wilks' lambda = .833364 $F(2, 40) = 3.999$ $p = .0261$

MANOVA for the hypothesis of no JA effect

Wilks' lambda = .1382491 $F(4, 80) = 33.790$ $p = .0001$

MANOVA for the hypothesis of no $JY1$ effect

Wilks' lambda = .894378 $F(2, 40) = 2.362$ $p = .1073$

Tests for hypotheses for between subjects effects

Source	df	Type III SS	Mean square	F value	$PR > F$
$Y1$	1	2005.72311	2005.7231	124.60	.0001
A	2	2650.48192	1325.2409	82.32	.0001
$E = ADJ \cdot RS(A)$	41	660.01022	16.0978		
Univariate tests of hypotheses for within subject effects					
J	2	37.70253	18.85126	3.72	.0284
JA	4	988.86328	247.21582	48.79	.0001
$JY1$	2	24.23087	12.11543	2.39	.0979
$E = ADJ \cdot RSJ(A)$	82	415.50246	5.06710		

G–G Epsilon = .9126, H–F epsilon = 1.0228

should be trivial. Yet we see the MANOVA test of J in Table 3.7 yielded an $F_{2,41} = 52.704$. In Table 3.8 the same test yields an $F_{2,40} = 3.999$, which is not even significant at the .01 level. We do not expect this covariate to produce such a drastic change on *any* within-subject effect. Looking at the estimated cell means, we see very little change. How could such trivial changes produce such a drastic change on the main effect of J?

What the covariance is expected to do is reduce the error term on the between-subjects portions of the analysis. That it does, since we see now that $MS_{Eb} = MS_{RS(A)} = 16.0978$, which is about one fourth the size of MS_{Eb} of Table 3.7. However, the peculiar results on the main effect of J are so suspicious that we better run the data another way to check out this entire summary table.

The author then ran the data on a program (CNDATA) that would yield all indices needed for each data point so that we could run on SAS GLM without having to use the REPEATED statement. Thus all 135 data values ($Y2$ to $Y4$) were input so that each had the corresponding A-value, RS-value, and J-value ($J2$ to $J4$), or 405 index values total. This is what had to be done to any repeated-measure design prior to version 5 of SAS, which introduced the REPEATED statement. Now the SAS statements are:

```
PROC GLM; CLASS A J RS;
MODEL Y = Y1 A RS(A) J A*J /SOLUTION; MANOVA J ;
LSMEANS A J A*J /STDERR PDIFF; MANOVA A*J ;
TEST H = A E = RS(A);
```

This gives us a different summary table, where the J effects make more sense than in Table 3.8. There are several differences.

Actually Table 3.9 is a composite created from several different outputs. Note that the $RS(A)$ is entered explicitly in the model term, as are the J and AJ factors. Of course, the J and AJ are needed since they will not be included by any REPEATED J 3; command. The inclusion of the $RS(A)$ term is needed to yield MS_{Ew} as the residual error term. Thus we get both error terms $MS_{Eb} = MS_{RS(A)}$, and $MS_{Ew} = MS_{RSJ(A)}$ in this run. However, the use of $RS(A)$ causes PROC GLM to treat each subject as a separate factor, so our model becomes overdetermined and cannot yield the LSMEANS that are our desired adjusted means. To obtain these, we need another run, leaving off the $RS(A)$ term in our model statement. In this run, the residual MSE is a result of pooling MS_{Eb} and MS_{Ew}, but we can ignore that. Also note that we used a new command: TEST H = A E = RS(A). This was done to provide the proper

TABLE 3.9. Repeat of the Covariance Analysis of Table 3.8 Without Using the REPEATED Command

Adjusted cell means

	$Y2=J2$	$Y3=J3$	$Y4=J4$	$\bar{Y}_{j.}$
$A1$	52.8964	59.8297	66.4964	59.7409
$A2$	60.1077	60.4411	58.8411	59.7966
$A3$	49.9958	50.5292	50.5292	50.3514
$\bar{Y}_{.k}$	54.333	56.9333	58.6222	56.6296

MANOVA for the hypothesis of no J effect

Wilks' lambda = .5114147 $F(2, 84) = 40.125$ $p = .0001$

MANOVA for the hypothesis of no JA effect

Wilks' lambda = .3072416 $F(4, 84) = 47.350$ $p = .0001$

Tests for hypotheses for between subjects effects

Source	df	Type III SS	Mean square	F value	$PR > F$
$Y1$	1	2005.7231	2005.7231	124.60	.0001
A	2	2650.48192	1325.2409	82.32	.0001
$E = ADJ \cdot RS(A)$	41	660.01022	16.0978		

Univariate tests of hypotheses for within subject effects

Source	df	Type III SS	Mean square	F value	$PR > F$
J	2	420.10370	210.01851	40.13	.0001
JA	4	991.49629	247.87407	47.35	.0001
$E = ADJ \cdot RSJ(A)$	84	439.73333	5.23492		

MS ratio to test A, since the automatic $F = MS_A/MS_{\text{res}} = MS_A/MS_{Ew}$ is useless and must be ignored.

How else do Tables 3.8 and 3.9 differ? Note that the adjusted cell means are slightly different. This occurs because in Table 3.8 GLM did not really do a proper covariance using just one regression slope. Instead it did a covariance of $Y2$ ($J2$) as predicted by $Y1$ with a slope of 1.1753. Then GLM did a covariance with $Y1$ predicting $Y3$ ($J3$) with a slope of 1.1627. Then it did a covariance of $Y4$ ($J4$) as predicted by $Y1$ with a slope of .9163. These individual slopes were used to compute the predicted cell means. This can be confirmed by using the equation

$$\text{Adjusted } Y \text{ mean} = \bar{Y}_k - b[\bar{X}_k - \bar{X}.],$$

recalling that in the present case $Y1$ is the X or covariate predictor. Actually, the X means do not differ by very much, so they do not adjust the $Y2$–$Y4$ means by very much, but if you check them in Table 3.8, they do come from the preceding formula using the three slightly different *b*'s. This is not what covariance should be doing; it should be using a single *b* throughout. The single $b = 1.0848$, but you cannot get this when using the REPEATED J 3; command.

What else differs from Tables 3.8 and 3.9? Maybe it is easier to look at what is the same first. All the terms in the between-subjects effects are exactly the same. In the within-subjects effects, we see that the *JA* and $RSJ(A)$ terms are just about the same. They differ slightly because Table 3.8 included a $JY1$ term that is not included in Table 3.9. They could be made to match exactly if that term were forced into Table 3.9, but there is little need for that extra work.

The big difference (in fact, the one that alerted us to the inadequacy of Table 3.8) is in the tests of the *J*-effect. In Table 3.8 the MANOVA $F = 3.999$, whereas in Table 3.9 the MANOVA $F = 40.125$. Part of this is due to slight differences in error terms, but only the latter corresponds to the *J*-effects under every other analysis, and there is little doubt that SAS PROC GLM is doing something wrong when we use the REPEATED command in combination with a covariance. Similarly, the univariate test of *J* in Table 3.9 yields an *F* of 3.72 in Table 3.8 but a more believable *F* of 40.13 in Table 3.9. The reanalysis of the same data done in Table 3.9 convinces this author that covariance models on designs involving one or more repeated-measures factors should *not* be done using SAS PROC GLM in combination with the REPEATED command. Entering the data with all indexes for each value of *Y* is more work, but at present it seems to be necessary.

Let me comment that this is the first time I have ever detected a major "bug" in SAS. In the past I have found SAS to be the best mainframe statistics package for most analyses I have done, especially the analysis of covariance. There are other packages that will give you summary tables for the analysis of covariance, but they will not yield adjusted means. This is stupid; reporting a COV without adjusted means is like reporting an ANOVA without any means, or any MS_E. One package (one that I liked) in one case led me into a trap when it produced adjusted means when it should have reported (as SAS does) that they were nonestimable.

Thus it is with considerable surprise, and some discomfort, that I report the failure of SAS here, but there is no doubt that there is a failure. When you get into programs like GLM that are designed to handle an amazing variety of experimental designs, it is difficult to keep adding features without sometimes using a technique that will work some of the time, but not all. It is impossible to test an infinite number of possible designs in a finite period of time. Adding the feature of the REPEATED command makes GLM much easier to use in repeated-measures ANOVA. However, you must avoid this command if doing a covariance with repeated measures.

Let me comment on one danger of the computer. One of the unfortunate tendencies one sees is that many faculty and almost all students think that any analysis the computer prints out must be perfect. Those of us who have written computer programs know much better. We know that there are compromises and somewhat arbitrary decisions made in all programs. I personally consider SAS PROC GLM to be a hopeless program to use to expose intermediate results that can illustrate to graduate students the matrix algebra results that are being used, because GLM does not use the minimal number of orthogonal (if a balanced design) contrast vectors but instead uses a complex design matrix that always yields a singular $X'X$ matrix. Thus when you ask for the solution option, what you get is hard to explain to most persons who are not full-time statisticians. Thus we force researchers to operate by faith, and this faith may sometimes lead them down false paths.

One of the ways to encourage some computer sophistication is to encourage users to do several analyses of the same data set. What would have happened in this second example if the computer user had only run the analysis in Table 3.8. It would be easy to ignore the rather large differences in the J main effect means. A naive user might say "Well, they look different, but the computer says they are not different, so I guess it is just chance."

In reality, an adequate computer analysis should include some exploratory data analysis and some scatterplots if you are going to use covariance or correlation analyses. Searches for outliers and data-cleaning techniques are also worthwhile. I sometimes shudder when I think how many thesis results may be the result of some mispunched data point that should have a 09 but was punched as a 90 and never detected. Attention to variances also helps detect such anomalies, but most computer users attend only to means, even when the variances show more important trends. The average user seems to think of variances only with respect to assumptions that only add complexity to the analysis if rejected. Of course, the present article also is concentrating on means and adjusted means, but I hope it at least alerted the reader to the presence of VCV matrices and the fact that they often have useful interpretations. If SAS takes the time to correct the "bug" revealed in this chapter, it would be nice if they also fixed the program to automatically print out some of the

VCV matrices. With all the other automatic output, this would not increase the amount of paper output by that much.

When this data set was submitted to the computer, it was not the author's intention to discover a weakness of the SAS PROC GLM REPEATED command, but merely to illustrate its use. Attending to which computer outputs are reasonable and which are not reasonable was necessary to spot this bug. This should be the common attitude, not "What results did the great infallible computer give me?" but "Did this computer output really make sense?" Looking at many different computer outputs on the same data set should help in spotting difficulties.

3. Same Experiment with Three Different Criteria

One more data set seems appropriate for analysis. So far, we have had a choice between using univariate tests on the within-subject factors and MANOVA on these same factors. We should do at least one data set where MANOVA is forced on us. Such a data set is in Table 3.10 and includes a *U* variable and a *V* variable in addition to the *Y* variable in Table 3.4. Only the *Y* variable was obtained at the pretest, but the *Y*, *U*, and *V* variables were all obtained from each subject at the posttest, retention 1, and retention 2. We shall again use the pretest score as a covariate, not only for *Y* but also for *U* and *V*. In the *U* and *V* variables, it would not make sense to obtain a gain score, since the *U* and *V* variables are not on the same scale as the pretest. This is a common case, where the gain score option is not sensible.

Table 3.10 also illustrates the case where the *A*-, *RS*-, and *J*-values of the experimental variables are punched along with the *Y*-, *U*-, and *V*-values of the dependent variables. The pretest score has been relabeled as *X* to be consistent with conventional covariate notation. Notice that the same *X* is repeated three times for each subject, since each subject has but one pretest score. It is necessary for this to be present on each of the three occasions to do a covariance. This is a case where there is a single covariate value for each subject, such as in Winer (1971, pp. 796–805) and Games (1979, pp. 185–192). It is possible to have a covariance in which there is a different covariate value for each observation (Winer, 1971, pp. 805–809; Games, 1979, pp. 192–197), but we will not illustrate that here.

First, we will look at the MANOVA. On each effect we wish to test,

TABLE 3.10. $[RS_{15}wX$ in $(A_3)]*J_3w$ Y U V Multivariate Data with Covariate X

OBS	A	RS	J	Y	U	V	X	OBS	A	RS	J	Y	U	V	X
1	1	1	1	57	35	65	54	69	2	8	3	59	52	48	51
2	1	1	2	58	56	48	54	70	2	9	1	56	52	47	49
3	1	1	3	67	68	46	54	71	2	9	2	56	53	51	49
4	1	2	1	54	72	64	52	72	2	9	3	59	54	52	49
5	1	2	2	64	57	46	52	73	2	10	1	61	57	48	49
6	1	2	3	68	63	71	52	74	2	10	2	60	57	54	49
7	1	3	1	48	48	51	47	75	2	10	3	65	63	60	49
8	1	3	2	58	62	54	47	76	2	11	1	60	58	56	51
9	1	3	3	64	47	49	47	77	2	11	2	58	53	51	51
10	1	4	1	52	50	47	52	78	2	11	3	61	57	54	51
11	1	4	2	61	49	53	52	79	2	12	1	61	54	56	52
12	1	4	3	68	52	50	52	80	2	12	2	59	56	53	52
13	1	5	1	48	37	52	47	81	2	12	3	59	53	49	52
14	1	5	2	55	41	39	47	82	2	13	1	69	66	62	58
15	1	5	3	61	55	50	47	83	2	13	2	66	59	60	58
16	1	6	1	53	52	48	52	84	2	13	3	65	58	53	58
17	1	6	2	58	36	41	52	85	2	14	1	55	51	48	47
18	1	6	3	68	52	53	52	86	2	14	2	51	47	40	47
19	1	7	1	57	41	45	56	87	2	14	3	53	47	43	47
20	1	7	2	62	63	57	56	88	2	15	1	54	50	45	48
21	1	7	3	66	61	54	56	89	2	15	2	56	52	51	48
22	1	8	1	55	51	53	51	90	2	15	3	53	48	43	48
23	1	8	2	64	59	54	51	91	3	1	1	56	51	44	56
24	1	8	3	64	54	50	51	92	3	1	2	56	53	51	56
25	1	9	1	49	47	46	47	93	3	1	3	57	54	49	56
26	1	9	2	59	50	54	47	94	3	2	1	48	43	40	48
27	1	9	3	70	63	61	47	95	3	2	2	47	44	38	48
28	1	10	1	57	52	51	51	96	3	2	3	49	45	41	48
29	1	10	2	65	54	49	51	97	3	3	1	45	42	37	47
30	1	10	3	71	63	52	51	98	3	3	2	44	41	36	47
31	1	11	1	48	45	49	48	99	3	3	3	46	41	39	47
32	1	11	2	52	50	48	48	100	3	4	1	61	57	55	60
33	1	11	3	60	55	61	48	101	3	4	2	62	60	56	60
34	1	12	1	49	46	52	48	102	3	4	3	59	54	52	60
35	1	12	2	55	51	47	48	103	3	5	1	57	52	50	54
36	1	12	3	66	62	60	48	104	3	5	2	61	56	53	54
37	1	13	1	52	49	46	49	105	3	5	3	59	54	48	54
38	1	13	2	59	53	48	49	106	3	6	1	59	56	48	60
39	1	13	3	64	60	62	49	107	3	6	2	59	57	53	60
40	1	14	1	52	49	48	51	108	3	6	3	55	51	47	60
41	1	14	2	59	55	57	51	109	3	7	1	44	42	38	47

TABLE 3.10. Continued.

OBS	*A*	*RS*	*J*	*Y*	*U*	*V*	*X*	OBS	*A*	*RS*	*J*	*Y*	*U*	*V*	*X*
42	1	14	3	66	60	58	51	110	3	7	2	43	40	36	47
43	1	15	1	61	54	57	57	111	3	7	3	47	44	41	47
44	1	15	2	67	62	56	57	112	3	8	1	48	45	42	49
45	1	15	3	73	68	62	57	113	3	8	2	46	43	38	49
46	2	1	1	56	47	49	48	114	3	8	3	47	41	36	49
47	2	1	2	53	51	47	48	115	3	9	1	47	44	40	47
48	2	1	3	51	46	48	48	116	3	9	2	48	45	41	47
49	2	2	1	60	52	56	52	117	3	9	3	47	45	42	47
50	2	2	2	67	60	63	52	118	3	10	1	51	48	43	49
51	2	2	3	63	57	51	52	119	3	10	2	51	50	46	49
52	2	3	1	57	53	48	50	120	3	10	3	50	48	43	49
53	2	3	2	58	51	54	50	121	3	11	1	48	45	47	49
54	2	3	3	59	55	51	50	122	3	11	2	49	45	41	49
55	2	4	1	69	67	62	56	123	3	11	3	45	42	37	49
56	2	4	2	69	65	63	56	124	3	12	1	52	49	46	51
57	2	4	3	65	61	58	56	125	3	12	2	52	49	46	51
58	2	5	1	59	54	52	47	126	3	12	3	57	53	51	51
59	2	5	2	56	50	48	47	127	3	13	1	43	40	37	47
60	2	5	3	55	53	51	47	128	3	13	2	43	38	36	47
61	2	6	1	62	64	57	53	129	3	13	3	43	39	33	47
62	2	6	2	61	57	54	53	130	3	14	1	50	49	47	53
63	2	6	3	59	52	47	53	131	3	14	2	56	51	48	53
64	2	7	1	59	53	55	49	132	3	14	3	53	48	44	53
65	2	7	2	69	64	60	49	133	3	15	1	46	42	36	51
66	2	7	3	53	48	47	49	134	3	15	2	46	41	38	51
67	2	8	1	60	57	56	51	135	3	15	3	49	45	41	51
68	2	8	2	64	61	54	51								

we ask if this effect produced differences on *any* of the *Y*, *U*, and *V* variables. Now we already saw big effects on the *Y* variable, so we expect to find significance on the MANOVAs. Sometimes, if you had weak significant effects on *Y* and no effects on *U* and *V*, you might obtain nonsignificant MANOVAs. Power considerations hold for MANOVA as well as for ANOVA. Of course, the virtue of MANOVA is that it controls the risk of Type I error for the complete set of three criterion variables rather than just one. The cell means, marginal means for the *U* and *V* variables, and the MANOVA terms for the summary table are

TABLE 3.11. ANOVA and MANOVA Results on Multivariate Data Set of Table 3.10

	Variable U				Variable V			
	$J1$	$J2$	$J3$	$\bar{Y}_{k.}$	$J1$	$J2$	$J3$	$\bar{Y}_{k.}$
$A1$	48.533	53.200	58.867	53.533	51.600	50.067	55.933	52.533
$A2$	55.667	55.733	53.600	55.000	53.133	53.533	50.333	52.333
$A3$	47.000	47.533	46.933	47.156	43.333	43.800	42.933	43.356
$\bar{Y}_{.}$	50.400	52.156	53.133	51.896	49.356	49.133	49.733	49.407

MANOVA for the hypothesis of no overall a effect $[E = RS(A)]$

Wilks' lambda = .3285808 $F(6, 80) = 9.927$ $p < .0001$

MANOVA for the hypothesis of no overall J effect $[E = RSJ(A)]$

Wilks' lambda = .4555502 $F(6, 164) = 13.164$ $p < 0.0001$

MANOVA for the hypothesis of no overall AJ effect $[E = RSJ(A)]$

Wilks' lambda = .2602693 $F(12, 217) = 12.007$ $p < .0001$

E = Type III $SS\&CP$ matrix for: $RS(A)$

df = 42	Y	U	V
Y	2665.733	2627.711	2348.844
U	2627.711	3191.111	2756.377
V	2348.844	2756.377	2691.511

Note: Y means are not repeated since they are in Table 3.7.

given in Table 3.11. The commands needed to produce Table 3.11 follow:

```
PROC GLM; CLASS A J RS;
MODEL Y U V = A J A*J RS(A);
MANOVA H = A E = RS(A)/SHORT;
MANOVA H = J /SHORT;
MANOVA H = A*J /SHORT;
LSMEANS A J A*J;
MEANS A | J /TUKEY ALPHA = .01;
```

By including the $RS(A)$ term explicitly in the model, we find that the residual error will be $RSJ(A)$. This is the appropriate error term for the within-subject J and AJ effects, and this residual term is automatically used when no E term is included in the MANOVA statement. However, for the A effect, we must specify $E = RS(A)$ as the error term in the MANOVA statement. Note that we have used a new term $A \mid J$ in the means statement. This is an SAS convention to shorten such statements. $A \mid J$ is the same as including A J A*J in the statement. It thus asks for all means of the cells and marginal means involving the crossed factors A and J. We have used the full notation (prior to this) for clarity, and will continue to do so.

Table 3.11 is an ANOVA and MANOVA for the equal n case. We asked for the descriptive means with the MEANS AJ A*J/ TUKEY ALPHA = .01; and we asked for the LSMEANS also. These should agree in any ANOVA or MANOVA with cells with equal n's. However, when we do a covariance, we should no longer use the MEANS commands. Only LSMEANS can give us the adjusted means.

SAS commands needed to give a multivariate analysis of covariance follow:

```
PROC GLM; CLASS A J RS;
MODEL Y U V = X A J A*J RS(A);
MANOVA H = A E = RS(A)/SHORT;
MANOVA H = J /SHORT;
MANOVA H = A*J /SHORT;
LSMEANS A J A*J;
```

These commands lead to the results in Table 3.12. Actually you must do another run, leaving $RS(A)$ out of the model to obtain the adjusted cell means of Table 3.12. When the $RS(A)$ term is in the model, the cell means are reported as NON-EST.

You may note that there are no differences between Tables 3.11 and 3.12 on the within-subject effects. The results for J give us an $F_{6,164} =$ 13.164 in both tables. The results for the AJ effect give us an $F_{12,217} = 12.000$ in both tables. This is what we should expect, since the covariance can only influence between-subject factors when there is just one X per subject. On the multivariate test for the A effect, there is a big difference between Table 3.11 and Table 3.12. In Table 3.11 the converted $F_{6,80}$ is 9.27, whereas in Table 3.12 the same $F_{6,78}$ is 21.453. Again this occurs because there is a sizable correlation between the covariate X and the dependent measures Y, U, and V. In this case, the increase in precision is shown in an error matrix rather than in a single term, and we will not go into details about it. The error matrix appropriate for the multivariate test on A is given at the bottom of Table 3.11 for the MANOVA and at the bottom of Table 3.12 for the multivariate covariance. For each entry, the value in Table 3.12 is smaller than in Table 3.11, and this corresponds to smaller values of MS_E, as we

TABLE 3.12. Covariance and Multivariate Covariance Results on the Data Set of Table 3.10, Using X (Pretest Score) as the Covariate on All Three Variables.

Adjusted Cell and Marginal Means

	Variable U				Variable V			
	$J1$	$J2$	$J3$	$\bar{Y}_{k.}$	$J1$	$J2$	$J3$	$\bar{Y}_{k.}$
$A1$	48.625	53.292	58.959	53.625	51.683	50.150	56.016	52.616
$A2$	55.895	55.964	53.830	55.230	53.341	53.741	50.541	52.541
$A3$	46.678	47.211	46.611	46.833	43.043	43.510	42.643	43.065
$\bar{Y}$	50.400	52.156	53.133	51.896	49.356	49.133	49.733	49.407

MANOVA for the hypothesis of no A effect $[E = RS(A)]$
Wilks' lambda = .142042 $F(6, 78) = 21.493$ $p < .0001$

MANOVA for the hypothesis of no J effect $[E = \text{residual } E = RSJ(A)]$
Wilks' lambda = .4555502 $F(6, 164) = 13.164$ $p < .0001$

MANOVA for the hypothesis of no AJ effect $[E = \text{residual } E]$
Wilks' lambda = .2602693 $F(12, 217) = 12.007$ $p < .0001$

E = Type III $SS\&CP$ matrix for: $RS(A)$

df = 41	Y	U	V
Y	660.010	711.520	624.446
U	711.520	1360.456	1108.954
V	624.446	1108.954	1208.979

Note: Y adjusted means are not repeated, since they are in Table 3.9.

have seen previously. Multivariate MANOVA and multivariate covariance are exact extensions of univariate ANOVA and univariate covariance, except matrix algebra is used.

After a significant MANOVA, the best procedure would be to do a discriminant function to find the linear combination of the three variables that shows the greatest differences. However, in this case with three variables showing the same directional differences, many people would merely do a univariate analysis on each of the three variables. We have already done such an analysis on Y, and the same thing could be repeated for U and V. This would involve the same steps already illustrated for Y, and so will not be done here.

A final warning. If you are doing ANOVA by PROC GLM and you have equal n's, it may be more convenient to use the MEANS command rather than the LSMEANS command. The MEANS command has a far greater choice of multiple comparison procedures. Here we used the Tukey WSD (or HSD) in some of the output. Unfortunately, all of the multiple comparison procedures in the MEANS command are *not* robust to heterogeneous variances. SAS does *not* have any commands in PROC GLM that are robust to heterogeneous variances. Such procedures do exist for many of the special case analyses that can be done on GLM, but they are not general. This again is one of the trade-offs—GLM is indeed general, and because it is so general it cannot handle some problems as well as more specialized programs. To return to the main warning, however, when using GLM with unequal *n*'s or covariance always stick to LSMEANS. Despite the rather inadequate output that comes with LSMEANS A J A*J /STDERR PDIFF; you must use LSMEANS for such designs. MEANS merely computes descriptive means for the cells and then weights them with the cell *n*'s when computing marginal means. For unequal *n* cases, these marginal means do not correspond to the Type III *SS*'s you should be reporting in your summary table. Thus your means and your summary table do not correspond, and this is trouble. Similarly, when you use the analysis of covariance, you want not the simple descriptive means but the adjusted means. If in doubt, ask for both, and if the output from MEANS does not agree with the output of LSMEANS, throw away the MEANS output. Some people may take offense to this recommendation. It takes a rather long discussion to defend it in detail, so I will leave it as something you may have to take on faith.

In summary, we have dealt with repeated-measures designs. Initially we took a simple design, and considered several alternative ways of analysis, and how they relate. We then went to a more complex design, with one between-subject factor and one repeated-measures, or within-subject factor, with three or four levels. We have seen that in SAS PROC GLM there is a REPEATED command that simplifies input when doing an ANOVA. However, we have seen that this command produces faulty output if it is used in an analysis of covariance with repeated measures. Thus we had to have each factor explicitly labeled to do an analysis of covariance design with repeated measures. Doing such an analysis was illustrated first for the ANOVA case and then for the covariance case. Finally, we did a multivariate analysis of variance with one repeated-measure factor. We saw how to do MANOVAs with repeated measures. Finally, we did a multivariate covariance on a repeated-measures design.

Obviously, there are an infinite number of possible experimental designs. Hopefully, if you master the principles illustrated in the examples here, you could apply these skills to other designs. The world of computers now permits us to do analyses that could never be considered in the days when the hand calculator was our major computational tool. However, the common availability of computers permits naive users to do analyses they do not understand and cannot interpret properly. Perhaps we are reaching a stage of complexity where many research projects should have a statistician on the staff so that such complex analyses can be verified by an expert, and only reasonable interpretations printed.

References

Box, G. P. E. (1954). "Some theorems on quadratic forms applied in the study of analysis of variance problems. II. Effect of inequality of variance and correlation of errors in the two-way classification," *Annals of Mathematical Statistics*. **25,** 484–498.

Cochran, W. G. (1957). "Analysis of covariance: its nature and uses," *Biometrics*. **13,** 261–281.

Collier, R. O., Baker, F., Mandeville, G. K., and Hayes, T. (1967). "Estimates of test size for several test procedures based on conventional variance ratios in the repeated measures design," *Psychometrika*. **32,** 339–354.

Davidson, M. L. (1972). "Univariate versus multivariate tests in repeated measures experiments," *Psychological Bulletin*. **77,** 466–452.

Duncan, D. B. (1955). "Multiple range and multiple F tests," *Biometrics*. **11,** 1–42.

Elashoff, J. D. (1969). "Analyses of covariance: a delicate instrument," *American Educational Research Journal*. **6,** 383–402.

Fleishman, E. A. (1960). "Abilities at different stages of practice in rotary pursuit performance," *Journal of Experimental Psychology*. **60,** 126–171.

Games, P. A. (1962). "A factorial analysis of verbal learning tasks," *Journal of Experimental Psychology*. **63,** 1–11.

Games, P. A. (1971). "Multiple comparisons on means," *American Educational Research Journal*. **8,** 531–565.

Games, P. A. (1975). "Computer programs for robust analyses in multifactor analysis of variance designs." *Educational and Psychological Measurement*. **35,** 147–152.

Games, P. A. (1976). "Limitations of the analysis of covariance in intact groups quasi-experimental designs," *Journal of Experimental Education*. **44,** 51–54.

Games, P. A. (1979). "Assessment and statistical control of subject variables in longitudinal designs." In J. R. Nesselroade, and P. B. Baltes, (Eds.),

Longitudinal research in the study of behavior and development (179–198). New York: Academic Press.

Games, P. A. (1981). "The ANOVR series: analyses for repeated measures designs with optional robust procedures when assumptions are violated," *The American Statistician.* **35**(2), 109–110.

Games, P. A., Grey, G. S., Herron, W. L., and Pitz, G. F. (1980). "ANOVR: Analysis of variance with repeated measures," *Behavioral Research Methods and Instrumentation.* **12,** 476.

Greenhouse, S. W., and Geisser, S. (1959). "On methods in the analysis of profile data," *Psychometrika.* **24,** 95–112.

Humphreys, L. G. (1960). "Investigations of the simplex,"*Psychometrika,* **25,** 213–323.

Huynh, H., and Feldt, L. S. (1970). "Conditions under which mean square ratios in repeated measurements designs have exact *F* distributions," *Journal of the American Statistical Association.* **65,** 1582–1589.

Jones, M. B. (1962). "Practice as a process of simplification," *Psychological Review.* **69,** 274–294.

Mardia, K. V. (1971). "The effect of nonnormality on some multivariate tests and robustness to nonnormality in the linear model," *Biometrika.* **58,** 101–121.

Maxwell, S. E. (1980). "Pairwise multiple comparisons in repeated measure designs," *Journal of Educational Statistics.* **5,** 269–287.

Miller, R. G., Jr. (1981). *Simultaneous statistical inference,* 2nd ed. New York: Springer-Verlag.

Mitzel, H. C., and Games, P. A. (1981). "Circularity and multiple comparisons in repeated measure designs," *The British Journal of Mathematical and Statistical Psychology.* **34,** 253–259.

Myers, J. L. (1979). *Fundamentals of experimental design,* 3rd ed. Boston, Mass.: Allyn and Bacon.

Olson, C. J. (1974). "Comparative robustness of six tests in multivariate analysis of variance, *Journal of the American Statistical Association.* **69,** 894–908.

Rogan, J. C., Keselman, H. J., and Mendoza, J. L. (1979). Analysis of repeated measurements," *British Journal of Mathematical and Statistical Psychology.* **32,** 269–286.

Rouanet, H., and Lepine, D. (1970). "Comparison between treatments in a repeated-measures design: ANOVA and multivariate methods." *The British Journal of Mathematical and Statistical Psychology.* **23,** 147–163.

Satterthwaite, F. E. (1946). "An approximate distribution of estimates of variance components," *Biometrics Bulletin (superseded by Biometrics).* **2,** 110–114.

Schatzoff, M. (1966). "Sensitivity comparisons among tests of the general linear hypotheses," *Journal of the American Statistical Association.* **61,** 415–435.

Winer, B. J. (1971). *Statistical principles in experimental design* 2nd ed. New York: McGraw-Hill.

II Structuring Change

Chapter 4 Longitudinal Factor Analysis

JOHN TISAK

Department of Psychology
Bowling Green State University
Bowling Green, Ohio

WILLIAM MEREDITH

Department of Psychology
University of California, Berkeley
Berkeley, California

Abstract

There has been increasing interest in dynamic modeling, that is, the modeling of longitudinal data. When the number of variates repeatedly measured on few occasions is large, it is often useful to somehow summarize or condense these variates. This chapter describes, extends, and integrates one group of techniques—current methods of factor analysis—to the problem of parsimoniously analyzing longitudinal data. This covers the continuum from exploratory to confirmatory methods. The latter extreme encompasses what Jöreskog and Sörbom (1984) have labeled the measurement model. With the objective of "measuring the same thing" at different times, special constraints must be placed on the regression weights or the factor pattern matrix. This structure is discussed under the rubrics of stationarity and invariance. Additional concern is given to the problem of correlated unique factors and factorial collapse. Some of the espoused models also include multiple populations and period/practice effects. Finally, all the suggested methods adhere to current statistical standards, such as, maximum likelihood estimation and asymptotic χ^2-tests.

STATISTICAL METHODS IN
LONGITUDINAL RESEARCH, VOLUME I:
Principles and Structuring

1. Introduction

Recently, there has been a gradual shift away from what might be called static modeling, (i.e., observations, with accompanying theory, describing a single time point) to what is currently called dynamic modeling (i.e., observations repeatedly measured and the theory to explain these observations; Plewis, 1985). This change has generated the formulation of new techniques (cf. Nesselroade and Baltes, 1979) to describe and to explain the developmental process. Furthermore, the amount and availability of substantive research has never been greater (cf. Verdonik and Sherrod, 1984). The purpose of this chapter is to describe, extend, and integrate current methods of factor analysis to longitudinal data.

When the number of variates is large and the number of occasions is small, then for parsimony it is useful to summarize or condense the variables in a reasonable fashion. In this endeavor, two procedures are commonly used. (Other taxonomies or classifications of data reduction or measurement techniques are possible, but these methods, e.g., alpha factor analysis (Kaiser and Caffrey, 1965) have not been generalized to include multiple occasions and groups.) The first is more data descriptive and falls under the rubric of (principal) component analysis (Bentler, 1973; Kroonenberg, Lammers, and Stoop, 1985; Lohmöller, 1978; Millsap and Meredith, 1988; Tucker, 1963). Although this approach is important and useful, its description is beyond the scope of this presentation. Instead this chapter will be restricted to factor analysis; this latter method might be considered more statistical, that is, it is consistent with standard inferential statistics (Lawley and Maxwell, 1971). A further distinction or classification for this factor analytic approach is possible. If the procedure is highly structured, then it is usually labeled as confirmatory. On the other hand, when there is less knowledge about the data structure, the method to be employed should be more exploratory. Under a similar formulation, both of these approaches will be described in this chapter. Additionally, techniques, which might be considered in between exploration and confirmation, will be included.

It is important to note the limits of this longitudinal factor analytic model. One limitation is quite similar to what Jöreskog and Sörbom (1984) call the measurement model. Structural equation modeling (SEM), that is, structure imposed on the factors, will be excluded. With this distinction, longitudinal factor analysis might be viewed as a prerequisite or subset of SEM. However, even without this additional structure, factor scores and factor means may be obtained.

In an attempt to "define" or "idenify" the latent factors, the restrictions of stationarity and invariance will be introduced. The longitudinal factor analytic models will be expected to incorporate multiple populations and practice effects. Moreover, two ways of depicting change (i.e., change in factor means and factor associations) will be given. Finally, for the most part, these factor analytic models will make use of maximum likelihood estimation and asymptotic χ^2-tests, the current standards of statistical theory.

2. A Longitudinal Factor Analysis Model

In this section, a basic formulation of longitudinal factor analysis will be stated. However, even in this relatively simple form, this approach will be too general to be of much practical use. Thus, certain restrictions will be developed. Also, this section will contain a number of important extensions to the longitudinal factor analytic model.

At a particular time or occasion, let N individuals be measured on p variates. Let these individuals be indexed by $i \in \{1, 2, \ldots, N\}$ and let these variables be indexed by $j \in \{1, 2, \ldots, p\}$. Furthermore, assume that these *same* individuals are assessed on these *same* variables on each of m occasions. Let $s, t \in \{1, 2, \ldots, m\}$ index the occasion of measurement. Thus, the total number of observed variables is mp. For each occasion, let the common factors be indexed by $f \in \{1, 2, \ldots, r\}$. The total number of common factors will then be mr. In some cases the constraint of replications of identical variables may be relaxed, but the liberalization of this requirement is beyond the scope of this exposition and might better be pursued under the topic of test equating (Rock, 1982).

Next, the random variables and vectors will be specified. Define the $p \times 1$ random vector of manifest variables, $\mathbf{x}_s$, as $\mathbf{x}_s' = [X_{s1}, X_{s2}, \ldots, X_{sp}]$, where X_{sj} is the jth random variable that is observed at the sth occasion. It is on these manifest variates that the actual observations will be made; that is, these are the observed variables. Similarly, define $\mathbf{z}_s$, the $r \times 1$ common factor random vector, $\mathbf{z}_s' = [Z_{s1}, Z_{s2}, \ldots, Z_{sr}]$, where Z_{sf} is the fth random variable at the sth occasion. Define $\mathbf{u}_s$, the $p \times 1$ random vector of unique variates, to be $\mathbf{u}_s' = [U_{s1}, U_{s2}, \ldots, U_{sp}]$; U_{sj} is the jth random variable at the sth occasion. The common and unique factor variables are latent and are not observed directly. Finally, at the sth time, let $\mathbf{\Lambda}_s$ be a $p \times r$ factor pattern or regression weight matrix.

With these definitions and in matrix notation, a longitudinal factor analytic model may be written as

$$\mathbf{x}_s = \mathbf{\Lambda}_s \mathbf{z}_s + \mathbf{u}_s \tag{1}$$

and

$$\mathbf{\Sigma}_{st} = \mathbf{\Lambda}_s \mathbf{\Phi}_{st} \mathbf{\Lambda}_t' + \mathbf{\Psi}_{st}, \tag{2}$$

where, without loss of generality for our immediate purposes, the first moments may be written as $\mathscr{E}[\mathbf{x}_s] = 0$, $\mathscr{E}[\mathbf{z}_s] = 0$, and $\mathscr{E}[\mathbf{u}_s] = 0$, and where the second moments or, in this case, the covariances $\mathbf{\Sigma}_{st} = \mathscr{E}[\mathbf{x}_s, \mathbf{x}_t']$, $\mathbf{\Phi}_{st} = \mathscr{E}[\mathbf{z}_s, \mathbf{z}_t']$, and $\mathbf{\Psi}_{st} = \mathscr{E}[\mathbf{u}_s, \mathbf{u}_t']$, a diagonal matrix. $\mathscr{E}[\cdot]$ is the expectation operator. Notice that for a particular occasion (i.e., for the sth occasion) and only for that occasion (i.e., $s = t$), (1) and (2) are the standard factor analytic model (Harman, 1976; Mulaik, 1972).

In the preceding model are the implicit assumptions that the unique factors are uncorrelated with the common factors within and across occasions. The former requirement is usual in factor analysis. The latter is a strong assumption, but to weaken it one would need a more powerful substantive theoretical framework than is proposed here. Furthermore, there are also the implicit assumptions that the unique factors are uncorrelated with themselves within occasions, but are correlated with the same unique factor across occasions. Again, the first assumption is standard, but the second is reasonable since the unique factor is composed of a systematic component that is specific to a variate and an unsystematic or random component. Thus, one would expect nonzero covariances among the specific component for each variate across time.

Let $\mathbf{x}$, $\mathbf{z}$, $\mathbf{u}$, $\mathbf{\Sigma}$, $\mathbf{\Phi}$, and $\mathbf{\Psi}$ be partitioned vectors and matrices with the just-defined partitioned elements $\mathbf{x}_s$, $\mathbf{z}_s$, $\mathbf{u}_s$, $\mathbf{\Sigma}_{st}$, $\mathbf{\Phi}_{st}$, and $\mathbf{\Psi}_{st}$, respectively. Because of its structure , $\mathbf{\Psi}$ will be referred to as a *band diagonal* matrix—a symmetrical matrix whose submatrices are diagonal. Define $\mathrm{D}_{\mathbf{\Lambda}_b}$ as a *block diagonal* matrix,

$$\mathrm{D}_{\mathbf{\Lambda}_b} = \begin{bmatrix} \mathbf{\Lambda}_1 & 0 & \cdots & 0 \\ 0 & \mathbf{\Lambda}_2 & \cdots & 0 \\ \vdots & \vdots & \vdots & \vdots \\ 0 & 0 & \cdots & \mathbf{\Lambda}_m \end{bmatrix},$$

where the subscript b is used to indicate that the block diagonals $\mathbf{\Lambda}_s$ are not equal for all m.

Utilizing these partioned matrices, we may compactly write the fundamental equations of longitudinal factor analysis (LFA), stated in (1) and (2), as

$$\mathbf{x} = \mathrm{D}_{\boldsymbol{\Lambda}_b}\mathbf{z} + \mathbf{u} \tag{3}$$

and

$$\boldsymbol{\Sigma} = \mathrm{D}_{\boldsymbol{\Lambda}_b}\boldsymbol{\Phi}\mathrm{D}'_{\boldsymbol{\Lambda}_b} + \boldsymbol{\Psi}. \tag{4}$$

The factor pattern matrix, as defined in the longitudinal factor analytic model, stated explicitly in (3) and (4), is a block diagonal matrix. This is to say that for each occasion of measurement there are r and only r common factors at each time that are the predictors of the p manifest variables. In other words, the common factors at one time do not influence or predict the observed variables from another time.

2.1. *Stationarity*

As mentioned previously, even the above definition of the factor pattern matrix is too general. The condition of stationarity is an almost necessary requirement to make the model practical. Define the *stationary longitudinal factor analytic model* (SLFA) as (1) and (2) with $\boldsymbol{\Lambda}_s = \boldsymbol{\Lambda}$ or (3) and (4) with the factor pattern matrix

$$\mathrm{D}_{\boldsymbol{\Lambda}} = \mathrm{I} \otimes \boldsymbol{\Lambda},$$

where $\otimes$ denotes the Kronecker product (Graham, 1981).

For the basic univariate (Pedhazur, 1982) or multivariate (Morrison, 1976) regression models, the condition of stationarity is significant. In these situations, the question being raised is whether or not the relationships between the outcome and the predictor variates are the same for different time periods. However, unlike the regression situations already stated, for the factor-analytic model the predictor variables are unknown or latent variables. Hence, stationarity is crucial to ensure the identity of the factors across time. Under LFA the meaning of the factors may change with time. On the other hand, in SLFA the connotation of the factors remains invariant. Of course, lack of stationarity is substantively important because it is indicative of *a* change (i.e., a change in what one is measuring), but it also precludes meaningful longitudinal comparisons.

2.1.1. Extension for Multiple Populations. An important extension of the longitudinal factor analytic model is to multiple populations, groups, or cohorts. This extension allows one to consider changes to naturally occurring populations, such as gender, and to experimental conditions, such as treatment and control groups. Let these different populations be indexed by $k \in \{1, 2, \ldots, q\}$. Notice that a generalization to the number of occasions, m, is possible with multiple populations. That is, the measurement periods may vary with population. Hence, $s, t \in \{1, 2, \ldots, m_k\}$. Because this generalization is straightforward, and for notational simplicity, this possible enhancement will not undergo further elaboration.

If superscript (k) is used to indicate the observations for the kth population, the LFA model given by (3) and (4) may be rewritten to incorporate multiple populations as

$$\mathbf{x}^{(k)} = \mathrm{D}^{(k)}_{\mathbf{\Lambda}_b}\mathbf{z}^{(k)} + \mathbf{u}^{(k)} \tag{5}$$

and

$$\mathbf{\Sigma}^{(k)} = \mathrm{D}^{(k)}_{\mathbf{\Lambda}_b}\mathbf{\Phi}^{(k)}\mathrm{D}^{(k)\prime}_{\mathbf{\Lambda}_b} + \mathbf{\Psi}^{(k)}, \tag{6}$$

where the factor pattern matrix is given by

$$\mathrm{D}^{(k)}_{\mathbf{\Lambda}_b} = \begin{bmatrix} \mathbf{\Lambda}^{(k)}_1 & 0 & \cdots & 0 \\ 0 & \mathbf{\Lambda}^{(k)}_2 & \cdots & 0 \\ \vdots & \vdots & \vdots & \vdots \\ 0 & 0 & \cdots & \mathbf{\Lambda}^{(k)}_m \end{bmatrix}.$$

2.2. *Invariance*

Analogous to stationarity where constraints were place on the factor pattern matrix across time to facilitate comparisons, for multiple populations the factor pattern matrix is made invariant across populations. Using Lawley's selection theorem, Meredith (1964) demonstrated the fundamentals of invariance in factor analysis. Later with maximum likelihood techniques, Jöreskog (1971a) developed a statistical approach to this property. Invariance of the factor pattern matrix is defined when

$$\mathrm{D}^{(k)}_{\mathbf{\Lambda}_b} = \mathrm{D}_{\mathbf{\Lambda}_b}.$$

Equations (5) and (6) may then be written to reflect this restriction as

$$\mathbf{x}^{(k)} = \mathrm{D}_{\mathbf{\Lambda}_b}\mathbf{z}^{(k)} + \mathbf{u}^{(k)}$$

and

$$\mathbf{\Sigma}^{(k)} = \mathrm{D}_{\mathbf{\Lambda}_b}\mathbf{\Phi}^{(k)}\mathrm{D}_{\mathbf{\Lambda}_b} + \mathbf{\Psi}^{(k)}.$$

Again the purpose of invariance is to aid in the understanding of structure and change in different groups. The rubric ILFA (invariant longitudinal factor analytic) will be used to designate this model.

Parenthetically, this constraint in factor analysis is analogous to the one imposed in validity studies (Cleary, 1968) and in item response theory (Lord, 1980). Cleary (1968) defines a measure unbiased if the regression coefficients are equal in different groups. In item response theory, an item is considered unbiased in different populations if it has the same item response curve across these populations. With factor analysis in several populations, a set of variates may be considered unbiased if their relationship (functional form) to the latent variables is constant; that is, they have the same factor pattern matrix.

2.2.1. Special Case of Stationarity and Invariance. If the restrictions of both stationarity and invariance are utilized, then one has a very important case of the LFA model. In this situation the factor pattern matrix is given by

$$\mathrm{D}_{\boldsymbol{\Lambda}} = \mathrm{I} \otimes \boldsymbol{\Lambda}.$$

The stationary invariant longitudinal factor analytic (SILFA) model is important because it permits direct comparisons of change across both time and population. Furthermore, it maintains the same structure (i.e., the definition and relationship of the factors or latent variables) across these situations.

2.2.2. Extension for Retest Effects. In longitudinal research there is the ubiquitous problem of retest, practice, period, or chronological order effect (Lord and Novick, 1968). Simply because of repeated measurement, this effect induces a change in individuals' scores. Often the test scores increase due to practice. However, boredom or lack of motivation may influence the scores to decrease.

One way to model retest or period effects is to consider their function linear (Molenaar, 1985; Vinsonhaler and Meredith, 1966); in other words, at the item level, learning is a Markov process that leads to a linear change as a function of practice. However, since in the factor-analytic model the scores are centered about their means, the additive effect is subtracted out and only the multiplicative effect remains.

Extending (5) and (6), we propose the following model:

$$\mathbf{x}^{(k)} = \mathrm{D}_0^{(k)} \mathrm{D}_{\boldsymbol{\Lambda}_b}^{(k)} \mathbf{z}^{(k)} + \mathbf{u}^{(k)} \tag{7}$$

and

$$\boldsymbol{\Sigma}^{(k)} = \mathrm{D}_0^{(k)} \mathrm{D}_{\boldsymbol{\Lambda}_b}^{(k)} \boldsymbol{\Phi}^{(k)} \mathrm{D}_{\boldsymbol{\Lambda}_b}^{(k)\prime} \mathrm{D}_0^{(k)} + \boldsymbol{\Psi}^{(k)}, \tag{8}$$

where $\mathrm{D}_0^{(k)}$ is a diagonal matrix of multiplicative effects and where for notational simplicity $\mathrm{D}_0^{(k)}$ has been absorbed into $\mathbf{u}^{(k)}$ of (7) and $\boldsymbol{\Psi}^{(k)}$ of (8).

Notice the following considerations. First, $D_{\Lambda_b}^{(k)}$ must be one of the special cases described previously, or $D_0^{(k)}$ must be equal to the identity matrix I. Second, if $D_0^{(k)}$ is different for all k, then retest effects are confounded with any effect which might change the scale of measurement. Therefore, to consider retest, period, or chronological order effects, one should use cohort sequential designs (Schaie, 1965). Ideally $D_0^{(k)} = D_0$, a diagonal matrix, for all k. Finally, since retest effects can distort linear relationships, their omission will hamper any attempt of linear factor analysis. In summary, retest, period, practice, or chronological order effects must be suspected in all psychological longitudinal research. Therefore, these longitudinal models may profitably incorporate them. When these effects are included, the previous longitudinal models (SLFA, ILFA, and SILFA) are extended (E); hence, they will be labeled ESLFA, EILFA and ESILFA, respectively.

2.2.3. Extension for Factorial Collapse. Factorial collapse occurs when one or more factors are linear combinations of the remaining factors over time. If there is factorial collapse, then the factor covariance matrix $\mathbf{\Phi}^{(k)}$ will no longer be of full rank. Under such conditions, few computing algorithms will function. Hence, a reparametrization of $\mathbf{\Phi}^{(k)}$ is not only useful but necessary. To reparametrize, let

$$\mathbf{\Phi}^{(k)} = \mathbf{\Gamma}^{(k)}\mathbf{\Theta}^{(k)}\mathbf{\Gamma}^{(k)\prime}, \tag{9}$$

where the rank of $\mathbf{\Phi}^{(k)}$ equals the order of $\mathbf{\Theta}^{(k)}$, a diagonal matrix with positive diagonal elements, and the number of columns in $\mathbf{\Gamma}^{(k)}$. Of course, some other structure could be imposed on $\mathbf{\Phi}^{(k)}$. However, this is done in this section because this would lead into the area of SEM, which is beyond the scope of the topic of factor analysis. Factorial collapse is important when studying growth.

3. Distributional Assumptions

Throughout this chapter, it will usually be necessary to assume that the random vector $(\mathbf{x}', \mathbf{z}', \mathbf{u}')$ is jointly distributed as multivariate normal. This condition will then permit maximum likelihood estimation and asymptotic χ^2-tests (Jöreskog, 1967; Lawley and Maxwell, 1971). However, with recent developments (Bentler and Berkane, 1985; Browne, 1984), the multivariate normal requirement may be relaxed and generalized via general elliptical and arbitrary distribution theory.

4. Scaling

Unlike the single-time, single-population, ordinary factor-analytic situation, longitudinal factor analysis in multiple populations evokes the problem of scaling or units of measurement. In general, the cavalier and, unfortunately, prevalent practice of standardization of variates at each occasion and for each group destroys the scale of the measuring instrument. This in turn makes comparisons across time and populations impossible. How can one directly compare measurements with different units? Furthermore, as was demonstrated for factorial invariance (Meredith, 1964), one has little hope for obtaining a stationary and invariant factor pattern matrix. Thus, the very meaning or interpretation of the factors may become clouded.

This discussion prompts the question, Why standardize in the first place? If all measurements are commensurable (i.e., in the same units) then standardization is unnecessary. Unfortunately, this is seldom the case in psychology. Moreover, often the units are very disparate. This disparity in scale will then in turn affect the regression coefficients (i.e., the values in the factor pattern matrix). Hence, when the scaling is different, a "standardization" is necessary. Furthermore, to alleviate these problems, this scaling must be constant for all occasions and populations.

Notice than *any* constant rescaling will satisfy the issues raised. To pick *the* rescaling method, we need additional criteria. The following three rescaling techniques are suggested, although others are possible.

1. Choose a standard. For example, pick one population and one occasion and then rescale all other times and populations by the reciprocal of the standard deviations:

$$\mathrm{D}_{\text{rescale}} = [\mathrm{Diag}[\boldsymbol{\Sigma}_{ss}^{(k)}]]^{-1/2}$$

   for a particular population k and occasion s. Notice for the particular k and s that the rescaled matrix will be a correlation matrix, and, in general, all other rescaled matrices will not be a matrix of correlations. In essence, this makes the sth occasion and the kth population the standard for the units of measurement. If national or population norms or parameter estimates are available, then the reciprocal of these standard deviations may be used as the rescaling factor.
2. For each variate, choose a function of its maximum (or minimum)

variance across all populations and times of measurement:

$$D_{\text{rescale}} = [\text{Max}[\text{Diag}[\boldsymbol{\Sigma}_{ss}^{(k)}]]]^{-1/2}$$

for all s and k. This will represent the rescaled covariance matrices as quasi-correlations; that is, no diagonal element may be greater than 1 (however, some will be less than 1). This rescaling was used by Tisak and Meredith (1989).

3. In a generalization of Jöreskog (1971a), use the reciprocal square of the average variance:

$$D_{\text{rescale}} = \left[\text{Diag}\left[\frac{1}{m(N-q)} \sum_{k=1}^{q} (N_k - 1) \sum_{s=1}^{m} \boldsymbol{\Sigma}_{ss}^{(k)}\right]\right]^{-1/2},$$

where (10)

$$N = \sum_{k=1}^{q} N_k.$$

This will make the average of the rescaled covariance matrices a correlation matrix.

The intent of this list of procedures is to be suggestive rather than complete. The "best" approach should be tailored to the substantive questions addressed.

5. Estimation and Testing of the Longitudinal Factor-Analytic Model

In the previous sections, a longitudinal factor-analytic model was formulated, (i.e., the basic modeling equations were stated). Also, two important restrictions, stationarity and invariance, and three salient extensions, multiple populations, practice effects, and factorial collapse, were postulated. Since this exposition on factor analysis relies heavily on statistical theory, procedures for the estimation of parameters and the testing of the model will be presented shortly.

Although the boundary is becoming increasingly blurred, the categorization of exploratory and confirmatory analysis (McDonald, 1985; Mulaik, 1972) will be utilized in this presentation. In this regard, confirmatory factor analysis includes those situations where the definition or identity of the factors is entirely specified. This more structured approach is best when there is considerable knowledge of the dimensions of the phenomenon under study. In the other direction, if little is known about the facets of the research domain, then much less structure may be imposed. In this case, exploratory factor analysis is the preferred tool.

5.1. Exploration

Although very powerful techniques for confirmatory factor analysis exist (Bock and Bargmann, 1966; Jöreskog, 1969), they do require identification of the factors; that is, structure must be imposed on the factor pattern matrix that relates an observed variate to a latent variable. Often, this is referred to as a *binary identification constraint.* When this structure is not apparent, more exploratory methods are called for. Recently, Tisak and Meredith (1989) have proposed a procedure for exploratory longitudinal factor analysis. A major advantage of their technique is that the usual binary constraints are not required. Hence, a priori knowledge of the factors is not needed; the method determines the latent variables. Clearly, this is more in keeping with the original development and purpose of (exploratory) factor analysis (Harman, 1976; Mulaik, 1972; Thurstone, 1947).

Without a detailed elaboration, the key to the success of this method is a stagewise or conditional solution for sets of parameters—that is, in general, for $\{D_0^{(k)}\}$, $\{D_{\Lambda_b}^{(k)}\}$, $\{\mathbf{\Phi}^{(k)}$ or $\mathbf{\Gamma}^{(k)}\mathbf{\Theta}^{(k)}\mathbf{\Gamma}^{(k)'}\}$, and $\{\mathbf{\Psi}^{(k)}\}$. After initial or start values have been obtained for all the parameters, each set of parameters is estimated conditionally on the remaining sets of parameters. By judicially choosing the sets, we avoid indeterminacies and, hence, the need for identification. The procedure is iterative—estimating and reestimating the sets of parameters—until the multivariate normal log-likelihood function is maximized. The actual iterative procedure employed is a stagewise expectation maximization (EM) algorithm (Dempster, Laird, and Rubin, 1977; Rubin and Thayer, 1982; Tisak and Meredith, 1989).

This exploratory longitudinal factor analysis method incorporates all the features discussed previously. It will, of course, handle from 1 to m occasions and from 1 to q populations. Also permissible are practice effects and/or factorial collapse. This is to say that the model depicted by (7) and (8) is possible. Again, all of this is obtainable without the specification of the factors.

Hence, using this technique, maximum likelihood estimates of the parameters, $D_0^{(k)}$, $D_{\Lambda_b}^{(k)}$, $\mathbf{\Phi}^{(k)}$ or $\mathbf{\Gamma}^{(k)}\mathbf{\Theta}^{(k)}\mathbf{\Gamma}^{(k)'}$, and $\mathbf{\Psi}^{(k)}$ are obtained. Furthermore, the factor pattern matrices $\mathbf{\Lambda}_s^{(k)}$ or one of its special cases (e.g., stationarity and/or invariance) has the "flavor" of principal components. That is, the columns are orthogonal and initially generated to span the common factor space. As an aside, notice that since no explicit identification structure is imposed on this factor pattern matrix, then similar to traditional factor analysis, this matrix can be rotated to simple structure (Thurstone, 1947) by any oblique rotation procedure, such as orthoblique (Harris & Kaiser, 1964), to increase the interpretability of the factors.

Under the assumption of multivariate normality, which is assumed throughout this chapter, it is possible to statistically test the fit of any of the variety of LFA models that were proposed. Following generalized likelihood ratio principles, these tests will be asymptotically χ^2 tests (Mood, Graybill and Boas, 1974). Furthermore, nested models (e.g., stationarity and nonstationarity) may be contrasted (compared) by restricted or subtractive χ^2 tests (Fix, Hodges, and Lehmann, 1959).

Because these tests are sensitive to violations of the assumption of multivariate normality, an easy extension will be presented next to make them more robust. One class of distributions that extends the robustness of normal distributions is elliptical distributions. This class of distributions includes not only the multivariate normal distribution but also contains platykurtic and leptokurtic multivariate distributions. If the data have a nonzero multivariate kurtosis coefficient, then the following correction for kurtosis (Browne, 1984) may increase the fit of these models:

$$\chi^2_{\text{corrected}} = \eta^{-1}\chi^2,$$

where η is a rescaling of Mardia's (1970) sample measure of kurtosis:

$$\eta = N^{-1} \sum_{i=1}^{N} \frac{\{(\mathrm{X}_i - \bar{\mathrm{X}})'\mathrm{S}^{-1}(\mathrm{X}_i - \bar{\mathrm{X}})\}^2}{mp(mp+2)},$$

where X_i is an $mp \times 1$ vector of observations for the ith subject, and $\bar{\mathrm{X}}$ and S are the maximum likelihood estimates of the population mean and covariance matrix, respectively.

In summary, with the basic equations (7) and (8), exploratory longitudinal factor analysis in a multiple population is possible. Identification of the factors is not required. Assuming multivariate normality, we may obtain maximum likelihood estimation of the parameters and asymptotic χ^2 tests. However, notice that this estimation and testing are not specific to the exploratory technique; it will also be used in the next section on confirmatory factor analysis.

5.2. Confirmation

With the development of confirmatory factor analysis (Bock and Bargmann, 1966; Jöreskog, 1969), with the allowance for several populations

(Jöreskog, 1971a), and with the conceptualization of factor analysis as a measurement technique (Jöreskog, 1971b; Lord and Novick, 1968), the capability for a confirmatory longitudinal factor-analytic model was present. Recently in a number of papers (Jöreskog, 1979; Jöreskog and Sörbom, 1976, 1977, 1985), longitudinal factor analysis as a measurement model has usually been depicted as a special case of (8); that is,

$$\boldsymbol{\Sigma}^{(k)} = \mathrm{D}_{\boldsymbol{\Lambda}}\boldsymbol{\Phi}^{(k)}\mathrm{D}_{\boldsymbol{\Lambda}} + \boldsymbol{\Psi}^{(k)}.$$

Probably because this formulation was only part of a more elaborate structural equations model and because of limitations of LISREL VI (Jöreskog and Sörbom, 1984), the computer program used in the data analysis, the extensions of multiplicative period/practice effects, and factorial collapse were not considered. Notice that either practice effects or factorial collapse may be included in the program LISREL. Furthermore, this formulation easily generalizes to permit the testing of what we call stationarity and invariance.

Discursively, although the purpose of this chapter is to consider LFA and not SEM, the heuristics of SEM will be suggested. Longitudinal factor analysis can be conceptualized as a measurement model or a definition of constructs. On the other hand, SEM in longitudinal research might be considered the description and association of these constructs over time. Furthermore, unlike the LFA formulation, SEM requires the retention of the means of the variates (i.e., the variables may not be in deviation units). Currently, the two approaches to this problem are autoregression and generalized growth curves models of the latent variables (i.e., variables without measurement error). The former, first (or higher-order) autoregressive models, have been utilized extensively by Jöreskog (1979) and by Jöreskog and Sörbom (1977, 1985). Also, in a special autoregressive case, Swaminathan (1984) has proposed an LFA model that is somewhere between exploration and confirmation; that is, for this situation, the factor pattern matrix is not rotatable. Though much less common, the latter, generalized growth curves, is presented by Tisak and Meredith (Chapter 11 in Volume II). In summary, although the boundary is fuzzy, LFA may be viewed as a special case of SEM, that is, the situation where relational structure is not placed on the common factors. Thus, LFA's practical utility is when the research interest is on construct definition but not construct relationships.

We reiterate: Throughout this chapter we have tried to keep the distinction between the longitudinal factor analytic model, (7) and (8), and the various computer programs to analyze them. In the previous paragraph, LISREL VI (Jöreskog and Sörbom, 1984) was alluded to.

This computer program will perform longitudinal factor analysis (and a great many other things) by maximum likelihood estimation. Since it is confirmatory, however, specific identification constraints must be imposed upon the factor pattern matrix. Furthermore, this program will provide asymptotic χ^2 tests.

In addition to LISREL, which is the most widely used of the SEM computer programs, there are two other packages: EQS and COSAN. Each program has unique features. One interesting feature of EQS (Bentler, 1989) is that the underlying distribution need not be multivariate normal. Instead, EQS may make use of either elliptical distributions or even asymptotically distribution-free methods. Within the multivariate normal theory, COSAN (McDonald, 1978) performs maximum likelihood estimation and asymptotic testing. With properly identified models, it will fit LFA data. Furthermore, because it is very comprehensive, COSAN can handle all variations of (7) and (8).

5.3. Exploration through Confirmation and Vice Versa

At the beginning of this section, it was suggested that the distinction between confirmatory and exploratory analyses is not clear-cut. Thus, one may consider exploration through confirmation or confirmation through exploration. The former approach has been advocated by Kroonenberg and Lewis (1982). They suggest a series of "confirmatory" factor analyses to attempt an exploration of the data structure. However, notice that one could just as well proceed in the opposite direction. First perform an exploratory LFA and then use these results for confirmation; that is, the solution (parameter estimates) becomes a starting point for a more detailed confirmatory analysis.

6. The Study of Change in a Longitudinal Factor Analytic Model

As mentioned earlier, it is not the intent of this exposition to discuss structural equation models. Indeed models like those proposed by Jöreskog (1979), Jöreskog and Sörbom (1976, 1977, 1985), and Tisak and Meredith (Chapter 11 in this volume) provide a very important contribution to the understanding of longitudinal phenomena. However, LFA may be viewed as a intermediate stage or even a precursor to SEM. Hence, within the domain of LFA, two aspects—factor covariances and factor scores—of longitudinal change will be suggested.

6.1. *Factor Covariances and Correlations*

One consideration in the LFA model is the relationships among the latent variables or the factors and the changes in these relationships longitudinally. These associations are available in the factor covariance and correlation matrices. Some important features of these matrices that will be discussed subsequently are factorial relevancy and differentiation/ integration. Note that the presentation will be purely descriptive (i.e., no attempt to structure these facets will be made).

6.1.1. Factorial Relevancy. The importance or relevancy of a factor may be described by the variance of that factor. For example, as the factor's variance approaches zero, so does its ability to discriminate among the individuals. On the other hand, if a factor has a larger amount of variance when compared to other factors, then it is more important. From LFA, these variances may be obtained from the diagonal elements of the within occasion factor covariance matrices, $\mathrm{D}^{(k)}_{\mathbf{\Phi}_{ss}}$. When this property is compared for a particular factor across time, one is able to view its relevancy chronologically. This is to say that one compares the corresponding elements of $\mathrm{D}^{(k)}_{\mathbf{\Phi}_{ss}}$ with $\mathrm{D}^{(k)}_{\mathbf{\Phi}_{tt}}$ to depict variance changes across time.

6.1.2. Factorial Differentiation/Integration. Factorial differentiation describes a longitudinal process whereby the correlations or covariances among the factors become smaller (i.e., the factors become less related). When the factors become more related (i.e., the converse), then factorial integration occurs. Since the determinant of a correlation matrix ranges from 0, with some columns of the matrix linearly related, to 1, where the variates are uncorrelated, the generalized variance is a natural measure of factorial integration/differentiation (Olsson and Bergman, 1977). Let the correlation among the factors of the sth occasion for kth population be given by $\mathrm{R}^{(k)}_{ss} = \mathrm{D}^{-1/2}_{\mathbf{\Phi}^{(k)}_{ss}} \mathbf{\Phi}^{(k)}_{ss} \mathrm{D}^{-1/2}_{\mathbf{\Phi}^{(k)}_{ss}}$. Then $|\mathrm{R}^{(k)}_{ss}|$ for $s = 1, 2, \ldots, m$ gives the change in the factor correlations across time. As $|\mathrm{R}^{(k)}_{ss}|$ increases, the factors are becoming more differentiated; conversely, as $|\mathrm{R}^{(k)}_{ss}|$ decreases, the factors are becoming more integrated.

6.2. *Factor Means and Factor Scores*

In the discussion of the LFA model, it was assumed that the means of the variates were equal to zero. Although this assumption does nothing to impede the formulation of a measurement model (i.e., the estimation of parameter matrices and goodness of fit tests), it does not permit the description of change in either the factor scores or the factor means. Hence, in this section, the restrictions on the means will be relaxed. Let $\mathscr{E}[\mathbf{x}^{(k)}] = \boldsymbol{\mu}_{\mathbf{x}}^{(k)}$ and $\mathscr{E}[\mathbf{z}^{(k)}] = \boldsymbol{\mu}_{\mathbf{z}}^{(k)}$; notice that $\mathscr{E}[\mathbf{u}^{(k)}] = \mathbf{0}$, still. Define $\boldsymbol{\Xi}_{\mathbf{xx}'}^{(k)} = \mathscr{E}[\mathbf{x}^{(k)}\mathbf{x}^{(k)\prime}]$ and $\boldsymbol{\Xi}_{\mathbf{zz}'}^{(k)} = \mathscr{E}[\mathbf{z}^{(k)}\mathbf{z}^{(k)\prime}]$.

With these modifications, the first and second moments of (7) may be written as

$$\boldsymbol{\mu}_{\mathbf{x}}^{(k)} = \mathrm{D}_0^{(k)}\mathrm{D}_{\boldsymbol{\Lambda}_b}^{(k)}\boldsymbol{\mu}_{\mathbf{z}}^{(k)} \tag{11}$$

and

$$\boldsymbol{\Xi}_{\mathbf{xx}'}^{(k)} = \mathrm{D}_0^{(k)}\mathrm{D}_{\boldsymbol{\Lambda}_b}^{(k)}\boldsymbol{\Xi}_{\mathbf{zz}'}^{(k)}\mathrm{D}_{\boldsymbol{\Lambda}_b}^{(k)\prime}\mathrm{D}_0^{(k)} + \boldsymbol{\Psi}^{(k)}. \tag{12}$$

Equations (11) and (12) may be combined compactly as

$$\begin{bmatrix} \boldsymbol{\Xi}_{\mathbf{xx}'}^{(k)} & \boldsymbol{\mu}_{\mathbf{x}}^{(k)} \\ \boldsymbol{\mu}_{\mathbf{x}}^{(k)\prime} & 1 \end{bmatrix} = \begin{bmatrix} \mathrm{D}_0^{(k)} & \mathbf{0} \\ \mathbf{0}' & 1 \end{bmatrix} \begin{bmatrix} \mathrm{D}_{\boldsymbol{\Lambda}_b}^{(k)} & \mathbf{0} \\ \mathbf{0}' & 1 \end{bmatrix} \begin{bmatrix} \boldsymbol{\Xi}_{\mathbf{zz}'}^{(k)} & \boldsymbol{\mu}_{\mathrm{z}} \\ \boldsymbol{\mu}_{\mathbf{z}}^{(k)\prime} & 1 \end{bmatrix} \begin{bmatrix} \mathrm{D}_{\boldsymbol{\Lambda}_b}^{(k)\prime} & \mathbf{0} \\ \mathbf{0}' & 1 \end{bmatrix} \begin{bmatrix} \mathrm{D}_0^{(k)} & \mathbf{0} \\ \mathbf{0}' & 1 \end{bmatrix}$$
$$+ \begin{bmatrix} \boldsymbol{\Psi}^{(k)} & \mathbf{0} \\ \mathbf{0}' & 0 \end{bmatrix}. \tag{13}$$

As written in (13) and with any of the restrictions or constraints discussed previously, the LFA model with means may be analyzed by any of the confirmatory computer programs described earlier. Notice that the maximum likelihood estimates of the factor score means, $\boldsymbol{\mu}_{\mathbf{z}}^{(k)}$, can then be obtained.

Determination of the factor scores require more effort. As described in the appendix, a generalization of Thomson's (1951) method of regressing the latent variables on the manifest variables is possible. Thus, the factor scores may be determined by

$$\mathrm{Z}^{(k)} = \mathbf{1}_{N_k}\hat{\boldsymbol{\alpha}}' + \mathrm{X}^{(k)}\hat{\mathrm{B}}',$$

$$\hat{\boldsymbol{\alpha}} = \hat{\boldsymbol{\mu}}_z^{(*)} - \hat{\mathrm{B}}\hat{\boldsymbol{\mu}}_x^{(*)},$$

and

$$\hat{\mathrm{B}} = [\hat{\boldsymbol{\Xi}}_{\mathbf{zx}'}^{(*)} - \hat{\boldsymbol{\mu}}_{\mathbf{z}}^{(*)}\hat{\boldsymbol{\mu}}_{\mathbf{x}}^{(*)\prime}][\hat{\boldsymbol{\Xi}}_{\mathbf{xx}'}^{(*)} - \hat{\boldsymbol{\mu}}_{\mathbf{x}}^{(*)}\hat{\boldsymbol{\mu}}_{\mathbf{x}}^{(*)\prime}]^{-1}.$$

Estimates of the average of the first moments are obtained from

$$\hat{\boldsymbol{\mu}}_{\mathbf{x}}^{(*)} = \frac{1}{q}\sum_{k=1}^{q} \hat{\boldsymbol{\mu}}_{\mathbf{x}}^{(k)} = \frac{1}{q}\sum_{k=1}^{q} \hat{\mathrm{D}}_0^{(k)}\hat{\mathrm{D}}_{\boldsymbol{\Lambda}_b}^{(k)}\hat{\boldsymbol{\mu}}_{\mathbf{z}}^{(k)}$$

and

$$\hat{\boldsymbol{\mu}}_{\mathbf{z}}^{(*)} = \frac{1}{q}\sum_{k=1}^{q} \hat{\boldsymbol{\mu}}_{\mathbf{z}}^{(k)}.$$

Similarly, estimates of the average of the second moments are given by

$$\hat{\boldsymbol{\Xi}}^{(*)}_{\mathbf{xx}'} = \frac{1}{q}\sum_{k=1}^{q}\hat{\boldsymbol{\Xi}}^{(k)}_{\mathbf{xx}'} = \frac{1}{q}\sum_{k=1}^{q}[\hat{\mathrm{D}}_0^{(k)}\hat{\mathrm{D}}^{(k)}_{\boldsymbol{\Lambda}_b}\hat{\boldsymbol{\Xi}}^{(k)}_{\mathbf{zz}'}\hat{\mathrm{D}}^{(k)\prime}_{\boldsymbol{\Lambda}_b}\hat{\mathrm{D}}_0^{(k)} + \boldsymbol{\Psi}^{(k)}].$$

Notice that

$$\hat{\boldsymbol{\Xi}}^{(*)}_{\mathbf{zx}'} = \frac{1}{q}\sum_{k=1}^{q}\hat{\boldsymbol{\Xi}}^{(k)}_{\mathbf{zx}'} = \frac{1}{q}\sum_{k=1}^{q}\hat{\boldsymbol{\Xi}}^{(k)}_{\mathbf{xx}'}\hat{\mathrm{D}}^{(k)\prime}_{\boldsymbol{\Lambda}_b}\hat{\mathrm{D}}_0^{(k)}.$$

Maximum likelihood estimators of all these parameters, $\{\boldsymbol{\mu}_{\mathbf{z}}^{(k)}, \boldsymbol{\Xi}^{(k)}_{\mathbf{zz}'}, D_0^{(k)}, \mathrm{D}^{(k)}_{\boldsymbol{\Lambda}_b}, \boldsymbol{\Psi}^{(k)}\}$, are all available from a confirmatory factor analysis program. Finally, $\mathrm{X}^{(k)}$ is an $N_k \times mp$ matrix of observations on the manifest variables and $\mathbf{1}_{n_k}$ is an $N_k \times 1$ unit vector, while $\mathrm{Z}^{(*)}$ is an $N_k \times mp$ matrix of predicted factor scores.

A key feature of this method of calculating factor scores is that the additive constant, $\boldsymbol{\alpha}$, and the regression weights, B, are invariant across all populations. This ensures the same linear combination of manifest variates for all samples. Hence, population bias or distortion is eliminated.

7. An Illustration

This example utilizes longitudinal data[1] on ability measures that were collected at three time points (1970, 1971, 1972) for two groups (females and males). In particular, the number facility (NUMF), letter series (LETS), and number series (NUMS) subtests from Thurstone and Thurstone's (1962) Primary Mental Abilities Test (PMA) were used. Collectively these variates may be taken to represent Horn and Cattell's fluid intelligence (Cattell, 1963) or an index of reasoning ability. For a further description of these data refer to Nesselroade and Baltes (1974).

Although a number of birth-year cohorts were observed in the original study, only those children who were in eighth grade in 1970 are used in this example. The data were rescaled by (10), and the rescaling vector was [0.192, 0.247, 0.327]. The rescaled means and covariances for both groups are presented in Table 4.1.

Since the structure of these data is well known, a confirmatory longitudinal factor analysis was undertaken. In particular, (6) was restricted to a stationary and invariant factor pattern matrix for both

[1] The authors wish to thank John Nesselroade for permitting us to use these data.

TABLE 4.1. Rescaled Covariances and Means

Means			2.172	1.926	2.024	2.656	2.420	2.425	2.879	2.835	2.685
NUMF1	0.789		0.745	0.414	0.453	0.686	0.484	0.540	0.741	0.482	0.631
LETS1	0.520	1.057		0.857	0.417	0.562	0.658	0.512	0.715	0.453	0.616
NUMS1	0.511	0.568	0.892		0.901	0.527	0.516	0.592	0.617	0.495	0.629
NUMF2	0.704	0.502	0.556	1.037		1.056	0.747	0.684	0.905	0.596	0.824
LETS2	0.454	0.695	0.469	0.512	0.849		1.194	0.697	0.807	0.764	0.841
NUMS2	0.403	0.484	0.466	0.526	0.458	0.824		1.110	0.824	0.646	0.900
NUMF3	0.667	0.579	0.608	0.847	0.571	0.527	1.028		1.389	0.776	1.026
LETS3	0.459	0.762	0.460	0.528	0.734	0.488	0.577	1.040		1.020	0.823
NUMS3	0.462	0.592	0.534	0.580	0.568	0.588	0.631	0.641	1.015		1.339
Means	2.032	2.308	1.950	2.441	2.843	2.439	2.667	3.151	2.628		

Below the diagonal are the covariances and means for females ($N = 121$); above are these for males ($N = 93$).

groups:

$$\boldsymbol{\Sigma}^{(k)} = \mathbf{D}_{\Lambda}\boldsymbol{\Phi}^{(k)}\mathbf{D}_{\Lambda}' + \boldsymbol{\Psi}^{(k)}, \qquad k = 1, 2. \tag{14}$$

The rescaled covariance matrices were used as the input data to LISREL VI. The results of this analysis yield an excellent statistical fit: $\chi^2(40) = 44.45$ and $p = 0.290$. However, inspection of the correlations of the factors (i.e., $\hat{\rho}^{(k)} = \mathrm{D}_{\boldsymbol{\Phi}^{(k)}}^{-1/2}\boldsymbol{\Phi}^{(k)}\mathrm{D}_{\boldsymbol{\Phi}^{(k)}}^{-1/2}$) yielded

$$\hat{\rho}^{(1)} = \begin{bmatrix} 1.000 & 0.934 & 0.926 \\ 0.934 & 1.000 & 0.982 \\ 0.926 & 0.982 & 1.000 \end{bmatrix} \quad \text{and} \quad \hat{\rho}^{(2)} = \begin{bmatrix} 1.000 & 0.956 & 0.966 \\ 0.956 & 1.000 & 0.963 \\ 0.966 & 0.963 & 1.000 \end{bmatrix}.$$

The associated determinants were $|\hat{\rho}^{(1)}| = 0.004$ and $|\hat{\rho}^{(2)}| = 0.004$. These results, along with our knowledge of the phenomenon and with the computer program's warning that $\phi_{11}^{(k)}$ may not be identified (even though it was), lead us to suspect that factor collapse might be present.

With the possibility of singularity of the common factors, (14) was modified to incorporate (9) to give

$$\boldsymbol{\Sigma}^{(k)} = \mathrm{D}_{\Lambda}\boldsymbol{\Gamma}^{(k)}\boldsymbol{\Theta}^{(k)}\boldsymbol{\Gamma}^{(k)\prime}\mathrm{D}_{\Lambda}' + \boldsymbol{\Psi}^{(k)}, \qquad k = 1, 2.$$

In this example, $\boldsymbol{\Gamma}^{(k)}$ and $\boldsymbol{\Theta}^{(k)}$ are 3×2 and 2×2 matrices, respectively. Again, the purpose of this reparameterization of $\boldsymbol{\Phi}^{(k)}$ was to reduce its rank from 3 to 2. This model was also analyzed by LISREL VI, in which the surprising statistical result ($\chi^2 = 35.900$, $p = 0.735$) was obtained.

Because of multicollinearity, a more restricted (and nested) model had a smaller χ^2-value than a less restricted one!

The maximum likelihood estimates of $\mathbf{\Lambda}$, $\mathbf{\Gamma}^{(k)}$, and $\mathbf{\Theta}^{(k)}$ are

$$\hat{\mathbf{\Lambda}}' = [1.081 \quad 1.000 \quad 1.080],$$

$$\hat{\mathbf{\Gamma}}^{(1)} = \begin{bmatrix} 1.000 & 0.000 \\ 0.000 & 1.000 \\ -0.028 & 1.116 \end{bmatrix}, \qquad \hat{\mathbf{\Gamma}}^{(2)} = \begin{bmatrix} 1.000 & 0.000 \\ 0.000 & 1.000 \\ 1.510 & -0.007 \end{bmatrix},$$

$$\hat{\mathbf{\Theta}}^{(1)} = \begin{bmatrix} 0.450 & 0.419 \\ 0.419 & 0.452 \end{bmatrix} \quad \text{and} \quad \hat{\mathbf{\Theta}}^{(2)} = \begin{bmatrix} 0.355 & 0.457 \\ 0.457 & 0.634 \end{bmatrix}.$$

The estimates of the uniqueness are given in Table 4.2, with $\hat{\mathbf{\Psi}}^{(1)}$ below the diagonal and $\hat{\mathbf{\Psi}}^{(2)}$ above.

Finally, the correlations among the common factors,

$$\hat{\rho}^{(k)} = \mathbf{D}^{-1/2}_{\mathbf{\Gamma}^{(k)}\mathbf{\Theta}^{(k)}\mathbf{\Gamma}^{(k)\prime}} \mathbf{\Gamma}^{(k)}\mathbf{\Theta}^{(k)}\mathbf{\Gamma}^{(k)\prime} \mathbf{D}^{-1/2}_{\mathbf{\Gamma}^{(k)}\mathbf{\Theta}^{(k)}\mathbf{\Gamma}^{(k)\prime}},$$

are given as

$$\hat{\rho}^{(1)} = \begin{bmatrix} 1.000 & 0.929 & 0.926 \\ 0.929 & 1.000 & 1.000 \\ 0.926 & 1.000 & 1.000 \end{bmatrix} \quad \text{and} \quad \hat{\rho}^{(2)} = \begin{bmatrix} 1.000 & 0.963 & 1.000 \\ 0.963 & 1.000 & 0.963 \\ 1.000 & 0.963 & 1.000 \end{bmatrix},$$

where, of course, the rank of $\hat{\rho}^{(k)}$ is 2. One might further test if $\hat{\rho}^{(k)}$ is really a rank 1 matrix. This was tested via the restricted χ^2-test, which proved to be significant ($\chi^2_R(4) = 12.70$, $p = .013$).

In conclusion, a relatively simple longitudinal factor-analytic model was presented. The purpose was to demonstrate the principles of stationarity and invariance of the factor pattern matrix, factor collapse,

TABLE 4.2. Covariances of the Unique Factors

NUMF1	.33		.34	.0	.0	.17	.0	.0	.10	.0	.0
LETS1	.0	.49		.44	.0	.0	.13	.0	.0	−.07	.0
NUMS1	.0	.0	.34		.50	.0	.0	.08	.0	.0	−.01
NUMF2	.23	.0	.0	.47		.35	.0	.0	.09	.0	.0
LETS2	.0	.20	.0	.0	.37		.47	.0	.0	.12	.0
NUMS2	.0	.0	.00	.0	.0	.37		.42	.0	.0	.08
NUMF3	.15	.0	.0	.24	.0	.0	.37		.37	.0	.0
LETS3	.0	.25	.0	.0	.23	.0	.0	.51		.39	.0
NUMS3	.0	.0	−.01	.0	.0	.05	.0	.0	.38		.33

and statistical testing and estimation. With an observed level of significance of $p = .735$, more restrictions, such as equality of the unique factor variances and/or some structure on the factor means, would, in practice, have been explored.

Appendix. Determination of the Factor Scores

In a generalization of Thomson (1951), one may determine the factor scores in multiple populations by regressing the latent variables on the observed scores; that is,

$$\mathbf{z}^{(k)} = \boldsymbol{\alpha} + \mathrm{B}\mathbf{x}^{(k)} + \boldsymbol{\varepsilon}^{(k)},$$

where $\mathbf{z}^{(k)}$, $\mathbf{x}^{(k)}$, and $\boldsymbol{\varepsilon}^{(k)}$ are all random vectors with dimensions $mr \times 1$, $mp \times 1$, and $mr \times 1$, respectively. The vector of additive constants, $\boldsymbol{\alpha}$, and the matrix of regression weights, B, are invariant across the q populations. This invariance maintains the same transformation of the observed variates for predicting the factors across all populations.

The first moments may be written as $\mathscr{E}[\mathbf{z}^{(k)}] = \boldsymbol{\mu}_{\mathbf{z}}^{(k)}$, $\mathscr{E}[\mathbf{x}^{(k)}] = \boldsymbol{\mu}_{\mathbf{x}}^{(k)}$, and $\mathscr{E}[\boldsymbol{\varepsilon}^{(k)}] = \mathbf{0}$. The second moments may be defined as $\boldsymbol{\Xi}_{\mathbf{zz}'}^{(k)} = \mathscr{E}[\mathbf{z}^{(k)}\mathbf{z}^{(k)\prime}]$, $\boldsymbol{\Xi}_{\mathbf{xx}'}^{(k)} = \mathscr{E}[\mathbf{x}^{(k)}\mathbf{x}^{(k)\prime}]$, and $\boldsymbol{\Xi}_{\boldsymbol{\varepsilon\varepsilon}'}^{(k)} = \mathscr{E}[\boldsymbol{\varepsilon}^{(k)}\boldsymbol{\varepsilon}^{(k)\prime}]$. Finally, the residuals or errors, $\boldsymbol{\varepsilon}^{(k)}$, are assumed orthogonal to the predictors, $\mathbf{x}^{(k)}$, which is a usual regression assumption.

As is standard in regression analysis, choose $\boldsymbol{\alpha}$ and B to minimize the expectation of the sum over populations of the sum of squared errors over variables:

$$\min\left[\mathscr{E}\left[\sum_{k=1}^{q} \mathrm{tr}[\boldsymbol{\varepsilon}^{(k)}\boldsymbol{\varepsilon}^{(k)\prime}]\right]\right]$$
$$= \min\left[\mathscr{E}\left[\sum_{k=1}^{q} \mathrm{tr}[(\mathbf{z}^{(k)} - \boldsymbol{\alpha} - \mathrm{B}\mathbf{x}^{(k)})(\mathbf{z}^{(k)} - \boldsymbol{\alpha} - \mathbf{B}\mathbf{x}^{(k)})']\right]\right].$$

This function may be minimized by symbolic differentiation (Schönemann, 1985).

Differentiating first with respect to $\boldsymbol{\alpha}$ yields

$$\frac{\partial}{\partial \boldsymbol{\alpha}}\left[\mathscr{E}\left[\sum_{k=1}^{q} \mathrm{tr}[\boldsymbol{\varepsilon}^{(k)}\boldsymbol{\varepsilon}^{(k)\prime}]\right]\right] = 2q\boldsymbol{\alpha} + 2\mathrm{B}\sum_{k=1}^{q} \boldsymbol{\mu}_{\mathbf{x}}^{(k)} - \sum_{k=1}^{q} \boldsymbol{\mu}_{\mathbf{z}}^{(k)}.$$

Equating this to zero and solving for $\boldsymbol{\alpha}$, one obtains

$$\alpha = \boldsymbol{\mu}_{\mathbf{z}}^{(*)} - \mathrm{B}\boldsymbol{\mu}_{\mathbf{x}}^{(*)}, \tag{A1}$$

where

$$\boldsymbol{\mu}_{\mathbf{z}}^{(*)} = \frac{1}{q}\sum_{k=1}^{q} \boldsymbol{\mu}_{\mathbf{z}}^{(k)} \quad \text{and} \quad \boldsymbol{\mu}_{\mathbf{x}}^{(*)} = \frac{1}{q}\sum_{k=1}^{q} \boldsymbol{\mu}_{\mathbf{x}}^{(k)}.$$

Next differentiating with respect to B produces

$$\frac{\partial}{\partial \mathrm{B}}\left[\mathscr{E}\left[\sum_{k=1}^{q} \operatorname{tr}[\boldsymbol{\varepsilon}^{(k)}\boldsymbol{\varepsilon}^{(k)'}]\right]\right] = 2\mathrm{B}\sum_{k=1}^{q} \boldsymbol{\Xi}_{\mathbf{xx}'}^{(k)} + 2\alpha\sum_{k=1}^{q} \boldsymbol{\mu}_{\mathbf{x}}^{(k)'} - 2\sum_{k=1}^{q} \boldsymbol{\Xi}_{\mathbf{zx}'}^{(k)}.$$

Substituting the preceding estimator for $\boldsymbol{\alpha}$, equating to $\mathbf{0}$, and solving for B yield

$$\mathrm{B} = [\boldsymbol{\Xi}_{\mathbf{zx}'}^{(*)} - \boldsymbol{\mu}_{\mathbf{z}}^{(*)}\boldsymbol{\mu}_{\mathbf{x}}^{(*)'}][\boldsymbol{\Xi}_{\mathbf{xx}'}^{(*)} - \boldsymbol{\mu}_{\mathbf{x}}^{(*)}\boldsymbol{\mu}_{\mathbf{x}}^{(*)'}]^{-1}, \tag{A2}$$

where

$$\boldsymbol{\Xi}_{\mathbf{zx}'}^{(*)} = \frac{1}{q}\sum_{k=1}^{q} \boldsymbol{\Xi}_{\mathbf{zx}'}^{(k)} \quad \text{and} \quad \boldsymbol{\Xi}_{\mathbf{xx}'}^{(*)} = \frac{1}{q}\sum_{k=1}^{q} \boldsymbol{\Xi}_{\mathbf{xx}'}^{(k)}.$$

If $\boldsymbol{\Sigma}_{\mathbf{xx}'}^{(*)} = \boldsymbol{\Xi}_{\mathbf{xx}'}^{(*)} - \boldsymbol{\mu}_{\mathbf{x}}^{(*)}\boldsymbol{\mu}_{\mathbf{x}}^{(*)'}$ and $\boldsymbol{\Sigma}_{\mathbf{zx}'}^{(*)} = \boldsymbol{\Xi}_{\mathbf{zx}'}^{(*)} - \boldsymbol{\mu}_{\mathbf{z}}^{(*)}\boldsymbol{\mu}_{\mathbf{x}}^{(*)'}$, then (A2) becomes $\mathrm{B} = \boldsymbol{\Sigma}_{\mathbf{zx}'}^{(*)}\boldsymbol{\Sigma}_{\mathbf{xx}'}^{(*)-1}$, which is a "standard" form for the regression coefficients. Notice that $\boldsymbol{\mu}_{\mathbf{x}}^{(*)}$ and $\boldsymbol{\mu}_{\mathbf{z}}^{(*)}$ are the unweighted parent population means and that $\boldsymbol{\Sigma}_{\mathbf{xx}'}^{(*)}$ and $\boldsymbol{\Sigma}_{\mathbf{zx}'}^{(*)}$ are the unweighted parent population covariance matrices.

Let N_k be the number of observations from the kth population, and let the total number of observations be $N = \sum_{k=1}^{q} N_k$. The weighted first moments may then be written as

$$\boldsymbol{\mu}_{\mathbf{x}}^{(\#)} = \frac{1}{N}\sum_{k=1}^{q} N_k \boldsymbol{\mu}_{\mathbf{x}}^{(k)} \quad \text{and} \quad \boldsymbol{\mu}_{\mathbf{z}}^{(\#)} = \frac{1}{N}\sum_{k=1}^{q} N_k \boldsymbol{\mu}_{\mathbf{z}}^{(k)};$$

the weighted second moments may be given as

$$\boldsymbol{\Xi}_{\mathbf{xx}'}^{(\#)} = \frac{1}{N}\sum_{k=1}^{q} N_k \boldsymbol{\Xi}_{\mathbf{xx}'}^{(k)} \quad \text{and} \quad \boldsymbol{\Xi}_{\mathbf{zx}'}^{(\#)} = \frac{1}{N}\sum_{k=1}^{q} N_k \boldsymbol{\Xi}_{\mathbf{zx}'}^{(k)}.$$

Using these first and second moments, we obtain the weighted parent population covariances matrices $\boldsymbol{\Sigma}_{\mathbf{xx}'}^{(\#)} = \boldsymbol{\Xi}_{\mathbf{xx}'}^{(\#)} - \boldsymbol{\mu}_{\mathbf{x}}^{(\#)}\boldsymbol{\mu}_{\mathbf{x}}^{(\#)'}$ and $\boldsymbol{\Sigma}_{\mathbf{zx}'}^{(\#)} = \boldsymbol{\Xi}_{\mathbf{zx}'}^{(\#)} - \boldsymbol{\mu}_{\mathbf{z}}^{(\#)}\boldsymbol{\mu}_{\mathbf{x}}^{(\#)'}$. Finally, similar to (A1) and (A2), the weighted regression coefficients may be given as

$$\boldsymbol{\alpha}_W = \boldsymbol{\mu}_{\mathbf{z}}^{(\#)} - \mathrm{B}_w \boldsymbol{\mu}_{\mathbf{x}}^{(\#)} \tag{A3}$$

and

$$\mathrm{B}_w = [\boldsymbol{\Xi}_{\mathbf{zx}'}^{(\#)} - \boldsymbol{\mu}_{\mathbf{z}}^{(\#)}\boldsymbol{\mu}_{\mathbf{x}}^{(\#)'}][\boldsymbol{\Xi}_{\mathbf{xx}'}^{(\#)} - \boldsymbol{\mu}_{\mathbf{x}}^{(\#)}\boldsymbol{\mu}_{\mathbf{x}}^{(\#)'}]^{-1}. \tag{A4}$$

Or if (A4) is rewritten in its more common form, it becomes

$$\mathbf{B}_w = \boldsymbol{\Sigma}_{\mathbf{zx}'}^{(\#)} \boldsymbol{\Sigma}_{\mathbf{xx}'}^{(\#)-1}.$$

References

Bentler, P. M. (1973). "Assessment of developmental factor change at the individual and group level." In J. R. Nesselroade and P. B. Baltes (Eds.), *Life-span developmental psychology* (pp. 145–174). New York: Academic Press.

Bentler, P. M. (1989). *EQS structural equations program manual.* Los Angeles: BMDP Statistical Software, Inc.

Bentler, P. M., and Berkane, M. (1985). "Developments in the elliptical theory generalization of normal multivariate analysis." Paper presented at the Annual Meeting of the American Statistical Association, Las Vegas, Nevada.

Bock, R. D., and Bargmann, R. E. (1966). "Analysis of covariance structures," *Psychometrika.* **31,** 507–534.

Browne, M. W. (1984). "Asymptotically distribution free methods for the analysis of covariance structures," *British Journal of Mathematical and Statistical Psychology.* **37,** 62–83.

Cattell, R. B. (1963). "Theory of fluid and crystallized intelligence: A critical experiment," *Journal of Educational Psychology.* **54,** 1–22.

Cleary, T. A. (1968). "Test bias: Prediction of grades of negro and white students in integrated colleges," *Journal of Educational Measurement.* **5,** 115–124.

Dempster, A. P., Laird, N. M., and Rubin, D. B. (1977). "Maximum likelihood from incomplete data via the EM algorithm," *Journal of the Royal Statistical Society, Series B.* **39,** 1–38.

Fix, E., Hodges, J. L. and Lehmann, E. L. (1959). "The restricted χ^2 test." In *Studies in probability and statistics dedicated to Harold Cramér.* Stockholm: Almquist and Wiksell.

Graham, A. (1981). *Kronecker products and matrix calculus: with applications.* New York: Wiley.

Harman, H. H. (1976). *Modern factor analysis.* Chicago: University of Chicago Press.

Harris, C. W. and Kaiser, H. F. (1964). "Oblique factor analytic solutions by orthogonal transformations," *Psychometrika.* **29,** 347–362.

Jöreskog, K. G. (1967). "Some contributions to maximum likelihood factor analysis," *Psychometrika.* **32,** 443–482.

Jöreskog, K. G. (1969). "A general approach to confirmatory maximum likelihood factor analysis," *Psychometrika.* **34,** 183–202.

Jöreskog, K. G. (1971a). "Simultaneous factor analysis in several populations," *Psychometrika.* **36,** 409–426.

Jöreskog, K. G. (1971b). "Statistical analysis of sets of congeneric tests," *Psychometrika.* **36,** 109–133.

Jöreskog, K. G. (1979). "Statistical estimation of structural models in longitudinal-development investigations." In J. R. Nesselroade and P. B. Baltes (Eds.), *Longitudinal research in the study of behavior and development* (pp. 303–351). New York: Academic Press.

Jöreskog, K. G. and Sörbom, D. (1976). "Statistical models and methods for test-retest situations." In D. N. M. de Gruijter and L. J. Th. van der Kamp (Eds.), *Advances in psychological and educational measurement* (pp. 135–157). London: Wiley.

Jöreskog, K. G. and Sörbom, D. (1977). "Statistical models and methods for analysis of longitudinal data." In D. J. Aigner and A. S. Goldberger (Eds.), *Latent variables in socioeconomic models* (pp. 285–325). Amsterdam: North-Holland.

Jöreskog, K. G., and Sörbom, D. (1984). *LISREL VI: Analysis of linear structural relationships by the method of maximum likelihood.* Chicago: Scientific Software, Inc.

Jöreskog, K. G., and Sörbom, D. (1985). "Simultaneous analysis of longitudinal data from several cohorts." In W. M. Mason and S. E. Fienberg (Eds.), *Cohort analysis in social research* (pp. 323–341). New York: Springer-Verlag.

Kaiser, H. F., and Caffrey, J. (1965). "Alpha factor analysis," *Psychometrika.* **30,** 1–14.

Kroonenberg, P. M., Lammers, C. J., and Stoop, I. (1985). "Three-mode principal component analysis of multivariate longitudinal organizational data," *Sociological Methods and Review.* **14,** 99–136.

Kroonenberg, P. M., and Lewis, C. (1982). "Methodological issues in the search for a factor model: Exploration through confirmation," *Journal of Educational Statistics.* **7,** 69–89.

Lawley, D. N. and Maxwell, A. E. (1971). *Factor analysis as a statistical method.* London: Butterworths.

Lohmöller, J. B. (1978). "How longitudinal factor stability, continuity, differentiation, and integration are portrayed into the core matrix of three-mode factor analysis." Paper presented at the European Meeting on Psychometrics and Mathematical Psychology, Uppsala, Sweden.

Lord, F. M. (1980). *Applications or item response theory to practical testing problems.* Hillsdale, NJ.: Lawrence Erlbaum.

Lord, F. M. and Novick, M. R. (1968). *Statistical theories of mental test scores.* Menlo Park, Calif.: Addison-Wesley.

Mardia, K. V. (1970). "Measures of multivariate skewness and kurtosis with applications," *Biometrika.* **57,** 519–530.

McDonald, R. P. (1978). "A simple comprehensive model for the analysis of covariance structures," *British Journal of Mathematical and Statistical Psychology.* **31,** 59–72.

McDonald, R. P. (1985). *Factor analysis and related methods.* Hillsdale, N.J.: Lawrence Erlbaum.

Meredith, W. (1964). "Notes on factorial invariance," *Psychometrika.* **29,** 177–186.

Millsap, R. E., and Meredith, W. (1988). "Component analysis in cross-sectional and longitudinal data," *Psychometrika.* **53,** 123–134.

Molenaar, P. C. M. (1985). "A dynamic factor model for the analysis of multivariate time series," *Psychometrika.* **50,** 181–202.

Mood, A., M., Graybill, F. A., and Boas, D. C. (1974). *Introduction to the theory of statistics.* New York: McGraw-Hill.

Morrison, D. F. (1976). *Multivariate statistical methods.* New York. McGraw-Hill.

Mulaik, S. A. (1972). *The foundation of factor analysis.* New York: McGraw-Hill.

Nesselroade, J. R., and Baltes, P. B. (1974). "Adolescent personality development and historical change: 1970–1972," *Monograph of the Society for Research in Child Development.* **39**(1, Serial No. 154).

Nesselroade, J. R., and Baltes, P. B. (Eds.) (1979). *Longitudinal research in the study of behavior and development.* New York: Academic Press.

Olsson, U., and Bergman, L. R. (1977). "A longitudinal factor model for studying change in ability structure," *Journal of Multivariate Behavioral Research.* **12,** 221–241.

Pedhazur, E. J. (1982). *Multiple regression in behavioral research.* New York: Holt, Rinehart and Winston.

Plewis, I. (1985). *Analysing change.* New York: Wiley.

Rock, D. A. (1982). "Equating using the confirmatory factor analysis model," In P. W. Holland and D. B. Rubin (Eds.), *Test equating* (pp. 247–257). New York: Academic Press.

Rubin, D. B., and Thayer, D. T. (1982). "EM algorithms for ML factor analysis," *Psychometrika.* **47,** 69–76.

Schaie, K. W. (1965). "A general model for the study of developmental problems," *Psychological Bulletin.* **64,** 92–107.

Schönemann, P. H. (1985). "On the formal differentiation of traces and determinants," *Multivariate Behavioral Research.* **20,** 113–139.

Swaminathan, H. (1984). "Factor analysis of longitudinal data," In H. G. Law, C. W. Snyder, Jr., J. A. Hattie, and R. P. McDonald (Eds.), *Research methods for multimode data analysis* (pp. 308–332). New York: Praeger.

Thomson, G. H. (1951). *The factorial analysis of human ability.* London: University of London Press.

Thurstone, L. L. (1947). *Multiple factor analysis.* Chicago: University of Chicago Press.

Thurstone, L. L., and Thurstone, T. G. (1962). *SRA primary mental abilities.* Chicago: Science Research Associates.

Tisak, J., and Meredith, W. (1989). "Exploratory longitudinal factor analysis in multiple populations," *Psychometrika,* **54,** 261–281.

Tucker, L. R. (1963). "Implications of factor analysis of three-way matrices for measurement of change," In C. W. Harris (Ed.), *Problems in measuring change* (pp. 122–137). Madison, Wis.: University of Wisconsin Press.

Verdonik, F., and Sherrod, L. R. (1984). *An inventory of longitudinal research on childhood and adolescence.* New York: Social Science Research Council.

Vinsonhaler, J. F., and Meredith, W. (1966). "A stochastic model for repeated testing," *Multivariate Behavioral Research,* **1,** 461–477.

Chapter 5 Patterns of Change within Latent Variable Structural Equation Models

J. J. McArdle

Department of Psychology
The University of Virginia
Charlottesville, Virginia

Mark S. Aber

Department of Psychology
The University of Illinois
Champaign, Illinois

Abstract

This chapter demonstrates how selected hypotheses about patterns of change can be investigated using concepts derived from linear structural equation modeling. In Section 5.1 we discuss the concept of structuring correlations by structural equation methods. In Section 5.2 we describe some longitudinal data and illustrate technical aspects of various structural models for these summary statistics. Here we describe the relations between the algebraic and graphic representation of linear models and the nonlinear expectations they imply. In Section 5.3 four different models of change are presented and fitted to these longitudinal data using structural equation model algorithms. These models include an autoregressive change model, a difference components change model, a

* We thank our colleagues at the Jefferson Psychometric Lab for their assistance on this research, and we also thank John Horn. Bill Meredith, and John Nesselroade for the free use of their ideas.

† This research has been supported by grants from the National Institute on Aging (AG07137).

growth curve change model, and a factor analysis change model. In Section 5.4 some of these ideas are considered together in a multivariate framework. These examples demonstrate the merger of contemporary issues in developmental psychology and structural equation analysis.

1. Introduction

Many new methods and statistical techniques can be useful for the analysis of developmental data. These methods differ in their mathematical bases, statistical emphases, data organization, computer programs, and in their often colorful names. But these useful methods all share one common feature: Developmental methods are all focused on some understanding of *patterns of change in data.*

This chapter is a discussion about the analysis of patterns of change using *structural equation modeling* techniques (abbreviated as SEM). The principal features of SEM have been around since the turn of the century and have been developed in psychometrics, sociometrics, econometrics, and biometrics (see Cattell, 1966; McDonald, 1985; Loehlin, 1987). The principal features of the change analyses used here have been developed in all these areas as well (see Nesselroade, 1983; McArdle and Epstein, 1987). Another popular term is *linear structural relations,* or *LISREL,* which defines a well-known computer program for SEM analyses (in Joreskog and Sorbom, 1979, 1985). We use these structural models to describe alternative ways to *represent change within developmental statistics.*

In this presentation we first define the concept of a *structure* in structural equation modeling. This is presented in both algebraic and graphic detail. Next we define different structural patterns for change data, so our structural models reflect *dynamics.* The change concepts we examine are based on autoregression, differences, growth, and the general use of latent variables. We illustrate these models by applying each model to the same longitudinal data. The structural differences reflect different ideas about developmental change so they provide different results. These modeling alternatives have not always been clear to developmental researchers (for examples see Wohlwill, 1973, 1988). We think the clarity of the structural equation models can add clarity to the substantive ideas about change.

1.1. The Structure in Structural Equation Modeling

The key idea of a structure in a structural model was nicely expressed in the original definition of *path analysis* by Sewell Wright (1921):

> The ideal method of science is the study of the direct influence of one condition on another in experiments in which all other possible causes of variation are eliminated. Unfortunately, causes of variation often seem to be beyond control. In the biological sciences, especially, one often has to deal with a group of characteristics or conditions which are correlated because of a complex of interacting, uncontrollable, and often obscure causes. The degree of correlation between two variables can be calculated by well known methods, but when it is found it gives merely the resultant of all connecting paths of influence. . . . The present paper is an attempt to present a method of measuring the direct influence along each separate path in such a system and thus finding the degree to which variation of a given effect is determined by each particular cause. The method depends on the combination of knowledge of the degrees of correlation among the variables in a system with such knowledge as may be possessed of the causal relations. In cases in which the causal relations are uncertain the method can be used to find the logical consequences of any particular hypothesis in regard to them (p. 557).

Another key idea of SEM comes from the clear display of alternative hypotheses using path analysis diagrams (after Wright, 1921, 1934, 1984; McArdle, 1980). Figure 5.1 presents some path diagrams based on the first path analysis model presented by Sewell Wright in 1921 (and again in 1934). In this problem Wright was concerned with the causes of differences in the birth weight of guinea pigs. He suggested that birth weight came from a few underlying variables, including the gestational period and the rate of growth. Wright's first conceptual diagram is presented in Figure 5.1(a). Figure 5.1(b) gives Wright's (1934) path analysis of a set of correlations based on over 1300 guinea pig litters. He concluded, "that size of litter has more than three times as much linear effect on birth weight through the mediation of the effect on growth as through its effect on the length of gestational period, contrary to the results of Minot's verbal analysis" (1934, p. 179). This use of direct causal consequences is often a useful way to organize structural hypotheses about correlations or covariances.

Wright's first path model is redrawn here using some contemporary path analysis notation in Figure 5.1(c). These graphic displays are a visual description of a series of simultaneous equations. In direct causal models the initial flow of the arrows directs attention to the implicit causal hypotheses about the variables. This diagram also reflects a more

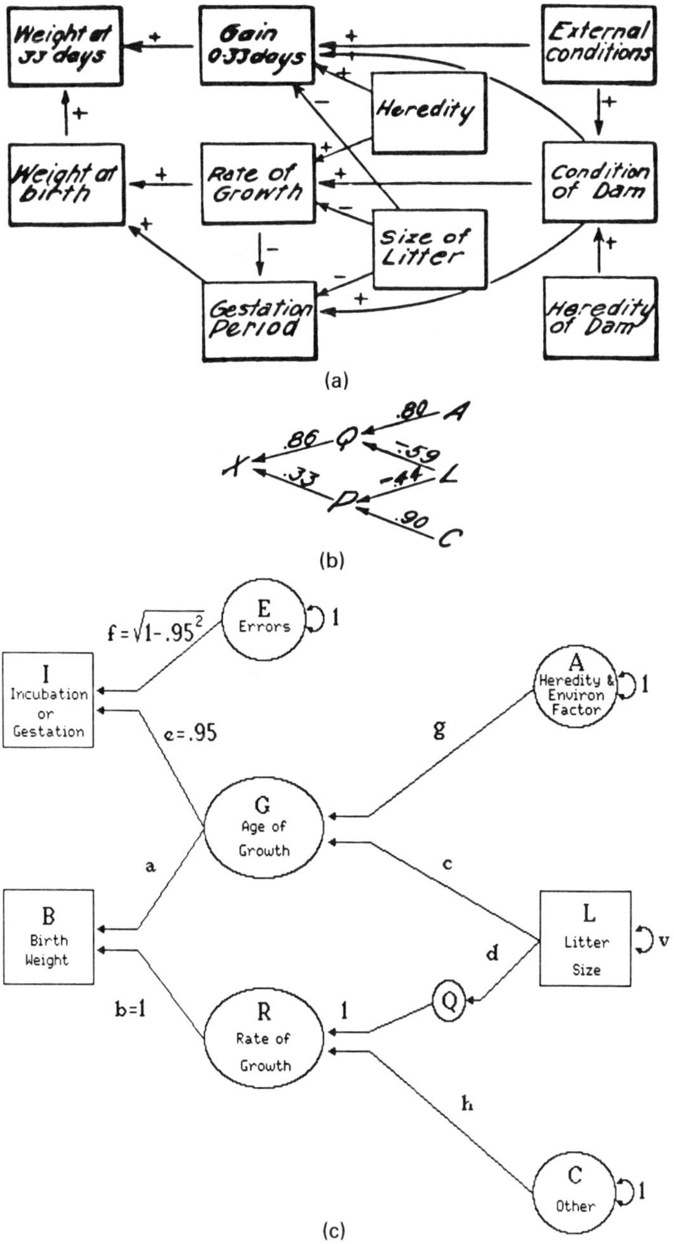

FIGURE 5.1. The first path analysis model: (a) theoretical description; (b) estimates from Minot's data (Figures 5.a and 5.b reprinted with permission of *The Annals of Mathematical Statistics*); (c) RAM structural diagram. (After Wright 1921, 1934).

complex set of interconnections which Wright (1934) called *the* .. *rules of path analysis*. These tracing rules are useful because they provide a visual description of the explicit structural hypotheses about the correlations. In all cases the relationship between the algebra and the graphics allows both to be better understood, and we demonstrate these issues here. This specific drawing (Figure 5.1(c)) also illuminates what is a remarkably contemporary example.

From Wright's work we see the *structure* in a structural equation model is a set of parameter-based hypotheses about patterns of relationships among the summary statistics. In the typical regression analysis we use several exogeneous X variables to maximize the explained variance R^2 of the dependent Y scores. As a by-product of this analysis we often obtain parameter estimates B_{yx} (of slopes or conditional means of Y as a function of means of X). By contrast, in SEM we try to predict the covariance (or the cross-products of the scores) of Y and X, and we are interested in the explained covariance RC^2. To test these structural hypotheses we form expected summary statistics from a model, we form observed summary statistics from data, and we compare these observations with these expectations.

1.2. *Latent Variables within Structural Models*

One of the useful aspects of SEM is the specification of models based on unobserved or *latent variables*. In direct causal models the specification of a measurement model allows the estimation and testing of causal hypotheses about reliable true scores or common factors of primary interest. This common factor approach is based on the logic originally described by Spearman (1904) in the first SEM (for details, see McDonald, 1985 or Loehlin, 1987).

Several latent variables were also indicated in the first path analysis models of Figure 5.1. The circles in Figure 5.1(c) are used to represent unobserved variables, but the latent variables G and R are the most critical circles here. In his definition of "age at growth" G, Wright implicitly made a formal separation of reliable and unreliable variables: This G is a hypothetical construct. In his definition of "rate of growth" R as a latent variable, however, Wright also defined latent variables without measurement parameters but with effects on the system: This R is a hypothetical process. In this remarkable first model Wright examined the equality of different paths of direct and indirect effects within a complex latent variable network with fixed and free parameters. (Incidentally,

Wright did not use equality of the covariance terms or the fitting of cross-product terms.) Wright's specific use of latent variables is hard to understand in his 1921 paper, but it is clarified in his 1934 version of the same analysis. It is not surprising to find path analysis has been misunderstood to some degree for almost 70 years.

Another broad goal for latent variable structural analyses has been nicely described by Nesselroade (1983):

> The identification of relationships that remain invariant among variables under different conditions and transformations is a major goal of empirical research. . . . Demonstration of factor invariance is one particular realization of a major goal of science—namely the identification of invariant relationships. The invariant relationships involved are those between factors (unobserved or latent variables) and observed variables or, in higher order analyses, other factors. At the first order of analysis for example, factor invariance signals a kind of constancy of a measurement system and thus the reasonableness of comparing phenomena in quantitative rather than in qualitative terms . . . From a change-measurement perspective, the emphasis in these models is on structural matters; establishing the properties of the measurement system and the nature of the variables first, then measuring and studying changes in them rather than on modeling and interpreting changes on observed variables directly. As such, factor invariance plays a crucial role in both the assessment of differences and change (pp. 59, 62–63).

In any structural regression we are primarily interested in the *invariance of parameters* B_{xy} over multiple replications. In latent variable analyses we examine the invariance of factor patterns and scores because we hope to refine our understanding of the most basic aspects of the constructs. Latent variable analysis is an indirect form of causal analysis, but it is certainly meaningful and effective. Our parameters are widely embedded in the hypothetical model and these indirect causal models communicate the structural hypotheses with great clarity and precision. In practice, we can use SEM to provide a structure for both direct and indirect causal models. These basic structural choices remain no matter whether we use correlations, covariances, cross-products, or other association matrices. We can embrace direct causal logic within our equations if warranted, or we can examine the invariance of relationships as a way to study outcomes of indirectly observed causal processes.

1.3. *Chapter Overview*

This chapter discusses questions about structural analysis of change data. We use Section 2 to describe some technical features of our structural equation approach. We first present some longitudinal data from previous experiments on intellectual ability in children by McArdle and Epstein (1987) and McArdle (1988). We also briefly explain the algebraic derivations needed for the model specification, and we discuss path graphic displays, such as Figure 5.1(c). We try to explain the algebraic and graphic relations in detail, but we do not detail statistical estimation or testing. We simply present an outline of how the specification of a model leads to the testing of structural hypotheses about change.

We use Section 3 to discuss a variety of different models of change. We start with *autoregressive models* because these use direct causal logic to form the key structural hypotheses. We explore a few uses of latent variables, including restricted sequences and the estimation of reliability coefficients. Second, we examine some *difference components models* of change. These models form indirect causal structures about developmental functions. Third, we explore some *growth curve models* of change for the same data set. Here we use the correlations, standard deviations, and the means to define some well-known models based on common patterns in means and covariances. These structural equation models include the traditional MANOVA (multivariate analysis of variance) models for repeated measures. All models are illustrated by fitting several alternative models to a small set of longitudinal data. Finally, we discuss *longitudinal factor analysis models* of change. This includes issues of changes in the number of factors, changes in the factor patterns, and changes in the factor scores.

In Section 4 we overview these models, describe some extensions, and define some limitations of our SEM approach. The limitations of SEM include the need for relatively high-quality data and the need for relatively strong developmental ideas. In theory, these are both features we would like to have in any experiment. In practice, both good data and good models are rarely ensured, and we must deal with whatever is available. For these reasons we must recognize SEM results are often unpredictable or in some cases unusable. In our view, however, these assumptions and problems stem from well-known problems of ANOVA, regression, and factor analysis. The problems and assumptions of SEM are more apparent because they are clearer.

In the appendix we add some technical information about mathematical and statistical model features and about computer programs. Throughout this presentation we recognize the current popularity of SEM is also due to the availability of new computer programs such as LISREL and COSAN. These programs and techniques allowed the formal specification of theory in terms of variables and their relations. In this presentation we show what these SEM programs can obtain, and we rely on these results (as in McDonald, 1985, or Loehlin, 1987). We illustrate some practical virtues and practical problems in using SEM techniques for change analyses.

This chapter is a basic overview of what we think are some key ideas in structural change analyses. Our structural approach here is similar to the presentation of Short, Horn, and McArdle (1983) and McArdle and Epstein (1987). The interested reader could also look at introductory treatments of Loehlin (1987) and McDonald (1985) or the more detailed articles of Joreskog and Sorbom (1979), Blalock (1985), or even McArdle (1986, 1988). Also, a variety of alternative and complementary approaches to change analyses are presented in the other chapters in this volume. This chapter is intended as a broad but introductory treatment on the structural equation modeling of developmental processes.

2. Methods

In this section we describe some technical features of SEM notation. These basic issues include data assumptions, algebraic notation, graphic notations, and algorithmic necessities. This is intended as a relatively basic introduction to the technical features of structural ideas and to our specific notation (so this section may be skipped by experts). The next section provides more details on methodological and substantive issues about different change models. Our main goal here is to introduce some data, calculate some statistics, and then to show how these models provide a structural prediction for these statistics.

2.1. Statistical Summary of Raw Data

Table 5.1 presents some summary data we can use here in model fitting. These data were collected in the experiment previously reported by Osborne and Suddick (1972), and aspects of this experiment have been presented in some detail by McArdle and Epstein (1987) and McArdle (1988). In these studies $N = 204$ children were measured on four occasions at approximately ages $A = \{6, 7, 9, \text{ and } 11\}$. The interval between these measurements is not equal, with age differences $dA = \{0, 1, 2, \text{ and } 2\}$, and this has an impact on our later structural models.

These children were repeatedly measured on a variety of psychological measurements. To simplify this presentation, we mainly use the *Wechsler*

TABLE 5.1. Summary Statistics for Osborne Longitudinal WISC Data (N = 204); see Osborne and Suddick, 1972; All WISC scores scaled as "percent correct of maximum."

(a) Means and standard deviations

	Occasion 1	Occasion 2	Occasion 3	Occasion 4
Age (in Years)	6.066	6.946	8.800	10.788
Means	18.034	25.819	35.255	46.593
Standard Dev.	6.374	7.319	7.796	10.386

(b) Correlation matrix

	WISC-1	WISC-2	WISC-3	WISC-4
WISC-1	1.000			
WISC-2	.809	1.000		
WISC-3	.806	.850	1.000	
WISC-4	.765	.831	.867	1.000

(c) Raw covariance matrix

	WISC-1	WISC-2	WISC-3	WISC-4
WISC-1	40.628			
WISC-2	37.741	53.568		
WISC-3	40.052	48.500	60.778	
WISC-4	50.643	63.169	70.200	107.869

(d) Scaled covariance matrix (scaled by first occasion standard deviation)

	WISC-1	WISC-2	WISC-3	WISC-4
WISC-1	1.000			
WISC-2	.929	1.319		
WISC-3	.986	1.194	1.496	
WISC-4	1.246	1.555	1.728	2.655

(e) Raw average cross-products matrix

	WISC-1	WISC-2	WISC-3	WISC-4	UNIT
WISC-1	365.661				
WISC-2	503.175	719.905			
WISC-3	675.656	958.479	1303.392		
WISC-4	890.680	1265.846	1712.475	2278.257	
UNIT	18.034	25.819	35.255	46.593	1.000

(*continued*)

TABLE 5.1. Continued.

(f) Scaled average cross-products matrix (note: values obtained by rescaling means and standard deviations by first occasion mean)

	WISC-1	WISC-2	WISC-3	WISC-4	UNIT
WISC-1	1.124				
WISC-2	1.547	2.213			
WISC-3	2.077	2.945	4.008		
WISC-4	2.739	3.892	5.265	7.001	
UNIT	1.000	1.432	1.955	2.584	1.000

Intelligence Scale for Children (WISC) scores, and we focus on the *WISC-verbal* subscale. This scale was created as the sum of Information, Comprehension, Similarities, and Vocabulary subscales. These scales were each transformed into *percent correct* units so that all new scores range between 0 and 100 and the mean gives the percent of items passed at each age. Table 5.1(a) gives means and standard deviations for the WISC-verbal score, and Table 5.1(b) gives the usual product-moment correlations among these WISC-verbal scores over time. The fact that precisely the same measurement device was used at each occasion also has an impact on our later structural models.

This kind of data summary is typical of most SEM experiments. But there is often a critical summary step that is overlooked: Every good SEM begins by an examination of univariate stem-and-leaf plots and bivariate and multivariate scatterplots (see Cleveland, 1985). Adequate distribution assumptions are required before we can be confident that all information in the scores is summarized by the means, deviations, and correlations. The features of nonlinearity, outliers, leverage points, linear relations, restriction of range, truncation and censoring, and other effects of selection are embedded within the raw scores. This is true for dependent or independent variables. If these scores do have any odd characteristics, these odd characteristics will be embedded in the summary statistics (see Belsley, Kuh, and Welsch, 1980). These fundamental problems of regression analyses tend to multiply problems for any SEM.

2.2. Correlations, Covariances, or Cross-Products?

The entries of the correlation matrix in Table 5.1(b) are averages of the mean deviated cross-products or multiplication of scores. Each statistic can be interpreted on an individual basis. For example, a standard Z-score of 1.5 on the WISC at time 1 and a standard Z-score of .5 on the WISC at time 2 produces a cross-product of .75 between occasions 1 and 2. The average of the multiplication of the time 1 and time 2 standard Z-scores taken over all people is .809 here. In contrast, Table 5.1(c) gives the covariance matrix for the four WISC scores. The entries of that matrix are average products of unstandardized but mean deviated scores. Again each statistic can be interpreted on an individual basis. For example, a mean deviate score of 5 on the WISC at time 1 and a mean deviate score of 8 on the WISC at time 2 produce a cross-product of 40 between occasions 1 and 2. The average of the multiplication of the time 1 and time 2 mean deviated scores over all people is 37.741 here.

The use of covariance matrices is often a confusing aspect of SEM analyses but the difference between correlations and covariances can be important. This covariance matrix is a product of the correlation matrix and the standard deviations over time. The covariances are necessary statistics because we are interested in any changes in the standard deviations over time. Covariances are important when models are used that employ *equality constraints* on parameters related to changes in the variances.

To illustrate this last point, we can rescale the covariance matrix so all standard deviations are scaled relative to an arbitrary reference point without any loss of statistical information. The matrix of Table 5.1(d) scales all deviations in relation to the size of the first occasion deviation, and, as a result, the scaled matrix looks much like a correlation matrix. But this scaled covariance matrix contains the same change information as the raw covariances (c), so we could use either for SEM. We recognize that most measurements in the behavioral sciences do not have the rigorous scaling systems available, say, in the physical sciences. So, while we are usually not interested in the absolute size of the variability over time, we are interested in the relative size of the changes in variability over time (e.g. the variance is 2.655 times larger at occasion 4).

Table 5.1(e) gives the product moment cross-products matrix for the four WISC scores. The entries in Table 5.1(e) are averages of the raw cross-products of scores, and each statistic can be interpreted directly. For example, a raw score of 20 on the WISC at time 1 and a raw score of 30 on the WISC at time 2 produce a cross-product of 600 between variables 1 and 2. The average of these cross-product scores over all people is 503.175 here. One additional variable is added to the data—a constant termed "UNIT" is assigned a value of 1 for all individuals. This constant is added to the data so, as in raw score regression, we can separate out intercept or location parameters based on the means.

This use of cross-products matrices seems even more confusing than our use of covariances, but it is motivated in the same way. The reason we use cross-products is that we are also interested in changes in the means. The difference between covariances and cross-products now becomes clearer: The cross-products matrix is a sum of the covariance matrix and the mean squares over time. Cross-products are necessary statistics because they retain the information about the changes in the means over time. This is especially important when models are used which employ *equality constraints* on parameters related to the means.

We would like to have ratio scales with equal intervals and a fixed zero point for all of our measurements. But in practice we are often limited to interval or ordinal scaling. To recognize this problem, we write our models to that the absolute size of the means does not usually matter and only the relative mean changes are critical. To illustrate this point statistically, we can rescale the cross-products matrix so that all means and deviations are scaled relative to a reference point without any loss of information. Table 5.1(f) scales all means and deviations in relation to the size of the first occasion mean. Again, this matrix looks more like a correlation or covariance matrix, but it contains the same change information as the raw cross-products (Table 5.1(e)) and we could use either for SEM.

2.3. Algebraic Model Specification

The specification of a structural equation model forms our primary definition of a model for change. In previous work we have introduced the use of an algebraic and graphical approach termed RAM (for details see McArdle and McDonald, 1984; McArdle and Horn, 1988; and the appendix). In brief, RAM notation is an extension of the traditional path diagram devices of Sewell Wright (1921, 1984) to allow more advanced models. In Table 5.2 we give an example of the algebraic formulation of one kind of model of change.

The equations in Table 5.2(a) gives the basic specification equation for a linear model. In this first model we assume a simple prediction of the score at some occasion $Y_{(t)}$ from the same score at the prior occasion $Y_{(t-1)}$. This prediction includes some unobserved error or disturbance terms $D(t)$. We also assume each variable has a nonzero coefficient labeled either a, b, c, h, j, or k. Table 5.2(b) adds some additional

TABLE 5.2 Algebraic Model Specification for Structural Path Models (See Figure 5.2)

(a) Internal (deterministic) model specifications

$$Y_{2(n)} = aY_{1(n)} + hD_{2(n)},$$
$$Y_{3(n)} = bY_{2(n)} + jD_{3(n)},$$
$$Y_{4(n)} = cY_{3(n)} + kD_{4(n)},$$

where for each subject n,

$Y_{t(n)}$ = observed score at occasion t

$D_{t(n)}$ = unobserved disturbances or errors of measurement at occasion t

a, b, c = regression coefficients

h, j, k = disturbance coefficients

(b) External (stochastic) model specifications

$$Y_{1(n)} = uY_{1(n)}, \quad \text{and} \quad \mathscr{E}\{uY_1 \quad uY_1'\} = z,$$
$$D_{2(n)} = uD_{2(n)}, \quad \text{and} \quad \mathscr{E}\{uD_2 \quad UD_2'\} = 1,$$
$$D_{3(n)} = uD_{3(n)}, \quad \text{and} \quad \mathscr{E}\{uD_3 \quad UD_3'\} = 1,$$
$$D_{4(n)} = uD_{4(n)}, \quad \text{and} \quad \mathscr{E}\{uD_4 \quad UD_4'\} = 1,$$

where

$uX_{(n)}$ = the external variable associated with internal variable $X_{(n)}$

$\mathscr{E}\{p \quad q'\}$ = the expected value of the deviations of variables p and q

z = the variance of Y_1

1 = the standardization of variables D_t required here for identification.

(c) Matrix model variables

$$v = \begin{bmatrix} Y_1 \\ Y_2 \\ Y_3 \\ Y_4 \\ D_2 \\ D_3 \\ D_4 \end{bmatrix}, \quad g = \begin{bmatrix} Y_1 \\ Y_2 \\ Y_3 \\ Y_4 \end{bmatrix}, \quad \text{and} \quad u = \begin{bmatrix} uY_1 \\ uY_2 \\ uY_3 \\ uY_4 \\ uD_2 \\ uD_3 \\ uD_4 \end{bmatrix}.$$

(d) Model parameter matrices

$$A = \begin{bmatrix} 0 & 0 & 0 & 0 & 0 & 0 & 0 \\ (a) & 0 & 0 & 0 & (h) & 0 & 0 \\ 0 & (b) & 0 & 0 & 0 & (j) & 0 \\ 0 & 0 & (c) & 0 & 0 & 0 & (k) \\ 0 & 0 & 0 & 0 & 0 & 0 & 0 \\ 0 & 0 & 0 & 0 & 0 & 0 & 0 \\ 0 & 0 & 0 & 0 & 0 & 0 & 0 \end{bmatrix}, \quad S = \begin{bmatrix} (z) & & & & & & \text{sym.} \\ 0 & 0 & & & & & \\ 0 & 0 & 0 & & & & \\ 0 & 0 & 0 & 0 & & & \\ 0 & 0 & 0 & 0 & (1) & & \\ 0 & 0 & 0 & 0 & 0 & (1) & \\ 0 & 0 & 0 & 0 & 0 & 0 & (1) \end{bmatrix},$$

$$F = \begin{bmatrix} (1) & 0 & 0 & 0 & 0 & 0 \\ 0 & (1) & 0 & 0 & 0 & 0 \\ 0 & 0 & (1) & 0 & 0 & 0 \\ 0 & 0 & 0 & (1) & 0 & 0 \end{bmatrix}.$$

components to this picture. This includes the assignment of one internal variable to each external variable (i.e., $Y_1 - uY_1$). These external variables (labeled u) represent variance contributed from all sources outside the model. The nonzero coefficients of Table 5.2(b) allow variable Y_1 to have variance z and require the unobserved disturbance variables to have unit variance.

The previous model has been called the *autoregressive simplex* model of change. This is a popular model, and it will be fitted to the WISC data in the next section. This model as stated in Table 5.2(a), (b) makes a rigid assumption that the disturbance terms $D(t)$ are uncorrelated with the predictors $Y_{(t-1)}$. In this model, this lack of correlation is an untestable assumption that is simply required for calculation of the other coefficients. Incidentally, this assumption is identical to the typical regression assumption of the independence of predictors and residuals. Still, this model forms a structural hypothesis that the pairwise correlation of $Y_{(t)}$ with $Y_{(t-1)}$ can be used to pattern all correlations among all $Y_{(t+x)}$. This is a testable assumption that can be formed by completing the matrix expectations (as described later).

Table 5.2(c) is a matrix specification for the model variables. In this representation all seven variables are ordered in some way in vector v, all four manifest variables are ordered in vector g, and the corrresponding seven external variables are ordered in vector u. Table 5.2(d) gives the corresponding matrix specification of the model coefficients.

The 7×7 matrix A of Table 5.2(d) defines the location of all asymmetric or deterministic regression-based coefficients of any model. These coefficients are entered into this matrix following the usual regression conventions—*column variables in v are input to row variables in v*. For example, the coefficient b is entered in the third row and second column because we assume variable $Y_{(3)}$ is an output of variable $Y_{(2)}$. In this matrix we enter all 6 non-zero coefficients embedded in the internal model (Table 5.2(a)).

The 7×7 matrix S defines the location of the symmetric or stochastic cross-product-based coefficients of any model. These coefficients are entered as cross-products so that column variables in v are associated with row variables in v. To be clear, the coefficient z is entered in the first row and first column because we assume variable $Y_{(1)}$ is associated with itself. In this matrix we enter all four nonzero coefficients embedded in the external model in Table 5.2(b).

The 4×7 matrix F is a set of fixed values which define the location of the manifest variables g within the full vector v. For example, the variable $Y_{(1)}$ is the first variable in g and the first variable in v, so we place a unit value in row 1 and column 1 of F. In this model we find the first four variables of g are located in the first four positions of v, so the F matrix is a partial identity and zero matrix.

These three matrices are important because they provide the *necessary and sufficient* matrices for any structural equation model (for proof, see McArdle and McDonald, 1984). This matrix specification is critical because it is usually required in computer programs for input and output. The appendix presents a more complete display of this general algebra.

2.4. Graphic Model Specification

The path diagrams of Figure 5.2 display the matrix equations of Table 5.2. In Figure 5.2(a) we draw the path diagram in the what has gradually

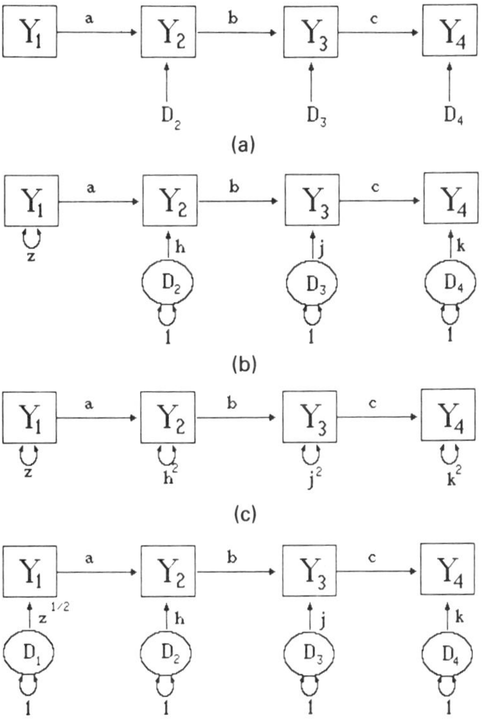

FIGURE 5.2. Equivalent structural path models for longitudinal data: (a) current path diagram; (b) complete RAM diagram; (c) compact but equivalent diagram, (d) expanded but equivalent diagram.

become a typical display (e.g., Joreskog and Sorbom, 1979, 1985). In these kinds of pictures we draw the four observed variables $Y_{(T)}$ as labeled squares, and we include the four unobserved disturbances as labels only. The best part of this diagram is the clear specification of the asymmetric coefficients drawn as one-headed arrows (from column toward row variable). Figure 5.2(b) redraws this model in a way that is consistent with the previous matrix algebra. In this picture we add all parameters and we distinguish all variables so that all structural predictions are clearly displayed. The graphics are clear, simple, and accurate.

In our view all linear structural equation models can be formed by making three main distinctions: (1) Elements are either *variables or parameters*; (2) Parameters are either *asymmetric or symmetric*; and (3) Variables are either *manifest or latent.* These three distinctions can be combined in many ways to form many different substantive models (see next section). But no other distinctions need be drawn in the graphics, algebra, or in the computer program (for more details see McArdle, 1980; Horn and McArdle, 1980; McArdle and McDonald, 1984; McArdle and Horn, 1988; McDonald, 1985; Loehlin, 1987).

All mathematical parameters are drawn in Figure 5.2(b). As usual, we include one-headed arrows for the asymmetric coefficients of matrix A. It is now relatively easy to see how matrix A of Table 5.2(d) is a matrix template for the arrows of Figure 5.2(b). We also include all elements of stochastic matrix S in Table 5.2(d) by adding two-headed arrows, here called *slings* or *spans,* to represent the nonzero elements of the S matrix. This includes one sling attached directly to variable $Y_{(1)}$ labeled z, and three other slings labeled 1 to represent the scaling of the disturbance variables $D_{(t)}$. These slings are unusual path graphic additions, but they are required for the complete graphic mapping of the elements of matrix S.

In a second break from tradition, we delineate all internal variables into two sets: All observed or manifest variables are drawn as squares, and all unobserved or latent variables are drawn as circles. This is done to allow a graphic location point for the required slings. This distinction clarifies the meaning of the nonzero elements of the F matrix—they indicate the squares. All internal variables in Table 5.2(a) are drawn in the picture, but the external u variables are never included in the picture. Again, the matrices are identical to the graphics, and vice versa.

The use of the nontraditional Figure 5.2(b) is more complete and useful than the common Figure 5.2(a). The key feature of RAM notation employed in Figure 5.2(b) is that the graphics provide a complete and entirely compatible representation of the previous algebra of Table 5.2. That is, the three matrices of Table 5.2(d) follow Figure 5.2(b) exactly. Our few changes closely follow Wright's original treatment, but they also allow us to completely match our algebra with our graphics. These additional features allow us to accurately portray the structural predictions. Perhaps more critically, these graphics allow us to generalize these principles to more complex models. These complete graphic parallels are not found in the path diagrams commonly displayed in the SEM literature, but we think they are useful.

2.5. *Algebraic Model Expectations*

We now present a simple but general model for the prediction of any structure of correlations, covariances, or cross-products (as presented by McArdle and McDonald, 1984; see the appendix). In this approach we first calculate the geometric series of matrix A. In Table 5.3(a) we show how the matrix A raised to the second power has two nonzero elements. The first element is ba, (b multiplied by a) and this is the length 2 arrow from row variable $Y(1)$ to column variable $Y(3)$. The second element is cb, (c multiplied by b) and this is the length 2 arrow from row variable $Y(2)$ to column variable $Y(4)$. These elements represent nonzero second-order effects of variables on other variables. Following this same logic, the matrix A raised to the third power has one nonzero element labeled abc from column variable $Y(1)$ to row variable $Y(4)$. The matrix A^4 is zero, and this means that there are no effects from one variable to another using four or more asymmetric coefficients in this model. The matrix of total effects E is formed as the sum of the nonzero powers of the A matrix. This calculation is the same for all linear path models.

The complete set of connections among all variables in the system is given by the product $C = ESE'$. In this example we obtain the C matrix of Table 5.3(b). The nonzero elements of this matrix are listed in terms of the basic parameters used to reproduce the association. Finally, we obtain the expected connections among the manifest variables $Cg_{ij'}$. This is created by filtering the rows and columns of the C matrix by calculating $C_g = FCF'$ as in the matrix equation in Table 5.3(c). In this example, the filter matrix simply serves to eliminate the disturbance variables $D_{(T)}$, but in later models all latent variable connections will be eliminated at this stage. This final matrix Cg reflects all of the model-based structural hypotheses of the observed statistics.

These structural hypotheses are often clarified by a more detailed look

TABLE 5.3. Algebraic Model Expectations for Structural Path Models*

(a) Expected effect calculation

$$A^2 = \begin{bmatrix} 0 & 0 & 0 & 0 & 0 & 0 & 0 \\ 0 & 0 & 0 & 0 & 0 & 0 & 0 \\ (ba) & 0 & 0 & 0 & 0 & 0 & 0 \\ 0 & (cb) & 0 & 0 & 0 & 0 & 0 \\ 0 & 0 & 0 & 0 & 0 & 0 & 0 \\ 0 & 0 & 0 & 0 & 0 & 0 & 0 \\ 0 & 0 & 0 & 0 & 0 & 0 & 0 \end{bmatrix}, \quad A^3 = \begin{bmatrix} 0 & 0 & 0 & 0 & 0 & 0 & 0 \\ 0 & 0 & 0 & 0 & 0 & 0 & 0 \\ 0 & 0 & 0 & 0 & 0 & 0 & 0 \\ (cba) & 0 & 0 & 0 & 0 & 0 & 0 \\ 0 & 0 & 0 & 0 & 0 & 0 & 0 \\ 0 & 0 & 0 & 0 & 0 & 0 & 0 \\ 0 & 0 & 0 & 0 & 0 & 0 & 0 \end{bmatrix}, \quad A^4 = 0,$$

$$E = A^0 + A^1 + A^2 + A^3 = \begin{bmatrix} 0 & 0 & 0 & 0 & 0 & 0 & 0 \\ (a) & 0 & 0 & 0 & (h) & 0 & 0 \\ (ba) & (b) & 0 & 0 & 0 & (j) & 0 \\ (cba) & (ca) & (c) & 0 & 0 & 0 & (k) \\ 0 & 0 & 0 & 0 & 0 & 0 & 0 \\ 0 & 0 & 0 & 0 & 0 & 0 & 0 \\ 0 & 0 & 0 & 0 & 0 & 0 & 0 \end{bmatrix}$$

(b) Expected summary matrices

$$C = ESE'$$

$$E = \begin{bmatrix} 0 & 0 & 0 & 0 & 0 & 0 & 0 \\ (a) & 0 & 0 & 0 & (h) & 0 & 0 \\ (ba) & (b) & 0 & 0 & 0 & (j) & 0 \\ (cba) & (ca) & (c) & 0 & 0 & 0 & (k) \\ 0 & 0 & 0 & 0 & 0 & 0 & 0 \\ 0 & 0 & 0 & 0 & 0 & 0 & 0 \\ 0 & 0 & 0 & 0 & 0 & 0 & 0 \end{bmatrix}$$

$$S = \begin{bmatrix} (z) & & & & & & \text{sym.} \\ 0 & 0 & & & & & \\ 0 & 0 & 0 & & & & \\ 0 & 0 & 0 & 0 & & & \\ 0 & 0 & 0 & 0 & (1) & & \\ 0 & 0 & 0 & 0 & 0 & (1) & \\ 0 & 0 & 0 & 0 & 0 & 0 & (1) \end{bmatrix}$$

$$E' = \begin{bmatrix} 0 & (a)' & (ba)' & (cba)' & 0 & 0 & 0 \\ 0 & 0 & (b)' & (ca)' & 0 & 0 & 0 \\ 0 & 0 & 0 & (c)' & 0 & 0 & 0 \\ 0 & 0 & 0 & 0 & 0 & 0 & 0 \\ 0 & (h)' & 0 & 0 & 0 & 0 & 0 \\ 0 & 0 & (j') & 0 & 0 & 0 & 0 \\ 0 & 0 & 0 & (k)' & 0 & 0 & 0 \end{bmatrix}$$

TABLE 5.3. Continued.

$$C = \begin{bmatrix} (C_{11}) & & & & & & \text{sym.} \\ (C_{21}) & (C_{22}) & & & & & \\ (C_{31}) & (C_{32}) & (C_{33}) & & & & \\ (C_{41}) & (C_{42}) & (C_{43}) & (C_{44}) & & & \\ 0 & 0 & 0 & 0 & (1) & & \\ 0 & 0 & 0 & 0 & 0 & (1) & \\ 0 & 0 & 0 & 0 & 0 & 0 & (1) \end{bmatrix},$$

where

$$\begin{aligned}
C_{21} &= az && = a(C_{11}) \\
C_{32} &= bh1h' + baza' && = b(C_{22}) \\
C_{43} &= cj1j' + cbh1h'b' + cbaza'b' && = c(C_{33}) \\
C_{31} &= bcz && = ab(C_{11}) \\
C_{42} &= abh1h' + cbaza' && = bc(C_{22}) \\
C_{41} &= cbaz && = abc(C_{11}) \\
C_{11} &= z && = z \\
C_{22} &= h1h' + aza' && = (aza' + hh') \\
C_{33} &= j1j' + bhh1h'b' + baza'b' && = b(aza' + hh')b' + jj' \\
C_{44} &= k1k' + cj1j'c' + cbh1h'b'c' + cbaza'b'c' && = c(b(aza' + jj')b')c' + kk'
\end{aligned}$$

(c) Manifest variable matrix equations

$Cg = FCF'$

$$= \begin{bmatrix} 1 & 0 & 0 & 0 & 0 & 0 & 0 \\ 0 & 1 & 0 & 0 & 0 & 0 & 0 \\ 0 & 0 & 1 & 0 & 0 & 0 & 0 \\ 0 & 0 & 0 & 1 & 0 & 0 & 0 \end{bmatrix} \begin{bmatrix} (C_{11}) & & & & & & \text{sym.} \\ (C_{21}) & (C_{22}) & & & & & \\ (C_{31}) & (C_{32}) & (C_{33}) & & & & \\ (C_{41}) & (C_{42}) & (C_{43}) & (C_{44}) & & & \\ 0 & 0 & 0 & 0 & (1) & & \\ 0 & 0 & 0 & 0 & 0 & (1) & \\ 0 & 0 & 0 & 0 & 0 & 0 & (1) \end{bmatrix} \begin{bmatrix} 1 & 0 & 0 & 0 \\ 0 & 1 & 0 & 0 \\ 0 & 0 & 1 & 0 \\ 0 & 0 & 0 & 1 \\ 0 & 0 & 0 & 0 \\ 0 & 0 & 0 & 0 \\ 0 & 0 & 0 & 0 \end{bmatrix}$$

$$= \begin{bmatrix} (C_{11}) & & & \text{sym.} \\ (C_{21}) & (C_{22}) & & \\ (C_{31}) & (C_{32}) & (C_{33}) & \\ (C_{41}) & (C_{42}) & (C_{43}) & (C_{44}) \end{bmatrix}$$

(d) Compact model parameter matrices

$$A = \begin{bmatrix} 0 & 0 & 0 & 0 \\ (a) & 0 & 0 & 0 \\ 0 & (b) & 0 & 0 \\ 0 & 0 & (c) & 0 \end{bmatrix}, \quad S = \begin{bmatrix} z & & & \text{sym.} \\ 0 & hh & & \\ 0 & 0 & jj & \\ 0 & 0 & 0 & kk \end{bmatrix}, \quad F = \begin{bmatrix} 1 & 0 & 0 & 0 \\ 0 & 1 & 0 & 0 \\ 0 & 0 & 1 & 0 \\ 0 & 0 & 0 & 1 \end{bmatrix},$$

so

$$E = \begin{bmatrix} 0 & 0 & 0 & 0 \\ (a) & 0 & 0 & 0 \\ (ba) & (b) & 0 & 0 \\ (cba) & (cb) & (c) & 0 \end{bmatrix}, \quad \text{and} \quad C = \begin{bmatrix} C_{11} & & & \text{sym.} \\ C_{21} & C_{22} & & \\ C_{31} & C_{32} & C_{33} & \\ C_{41} & C_{42} & C_{43} & C_{44} \end{bmatrix}.$$

* After McArdle and McDonald, 1984.

at the elementary scalar expressions. One easy way to see the model implications is to factor out the covariances C_{jj} and see the *reduced form* equations. When we do this, we find that the adjacent occasions have a simple connection given by a single arrow: $C_{12} = a(C_{11})$, $C_{23} = b(C_{22})$, and $C_{34} = c(C_{33})$. This gives us a way to estimate the coefficients from the data: that is, $a = (C_{12}/C_{11})$, $b = (C_{23}/C_{22})$, and $c = (C_{34}/C_{33})$. These estmates are equivalent to separate regressions. But this structure is more complex because we also make other restrictive predictions. If pairs of variables are two occasions apart, we have connections requiring two arrows: $C_{31} = ab(C_{11})$, and $C_{24} = bc(C_{22})$, so $ab = (C_{31}/C_{11})$ and $bc = (C_{24}/C_{22})$. Also, if the pairs are three occasions apart, we require three arrows: $C_{41} = abc(C_{11})$, so $abc = (C_{41}/C_{11})$. These are not part of the usual regression approach.

All structural restrictions are a result of the choice of a particular change model, but they are not a necessary feature of the summary statistics. From the first set of equations we assume $ab = (C_{12}/C_{11})(C_{23}/C_{22})$, but from the second set of equations we assume $ab = (C_{31}/C_{11})$. If the model is correct, these alternative calculations of a times b should both give the same answer. In essence, we predict $(C_{12}/C_{11})(C_{23}/C_{22}) = (C_{31}/C_{11})$, and we may be wrong. This predicted equality of a particular function of the correlations may be considered as the first structural restriction in the model. Here and in general we do not presume that any connection is zero or one, but we predict that one connection (C_{31}) is some function of other connections (C_{32} and C_{21}). The supposition that the connection between variables $Y(1)$ and $Y(3)$ is equal to the connection between $Y(1)$ and $Y(2)$ multiplied by $Y(2)$ and $Y(3)$ comes from our choice of this specific model.

Structural restrictions are never exactly true of any real data, so we add a statistical model (see later). In this approach we calculate the likelihood of this restriction in a theoretical population, and we obtain a single *degree of freedom* (DF) to test the adequacy of this restriction against another alternative. Two other restrictions come from the other additional constraints. From the first equations we assume $bc = (C_{23}/C_{22})(C_{34}/C_{33})$, and from the second we assume $bc = (C_{24}/C_{22})$, so we predict

$$(C_{23}/C_{22})(C_{34}/C_{33}) = (C_{24}/C_{22}).$$

Also, from the first equations we assume $abc = (C_{12}/C_{11})(C_{23}/C_{22})(C_{34}/C_{33})$, and from the third we assume $abc =$

(C_{41}/C_{11}), so we predict

$$(C_{12}/C_{11})(C_{23}/C_{22})(C_{34}/C_{33}) = (C_{41}/C_{11}).$$

Again, these restrictions are not necessarily true of the data, so we obtain a single DF to test this hypothesis. This collection of structural hypotheses leads to the overall DF for the model.

There are many simpler ways to enhance the patterning in these restrictions. By scaling all variances to $C_{jj} = 1$ we obtain a restriction where the product of two correlations is supposedly equal to the other (i.e., $C_{12}C_{23} = C_{31}$). This elementary approach is comparable to the tetrad difference approach used in the first SEM by Spearman (1904). We could also simplify the model further by assuming a single effect coefficient $p = a = b = c$. This would lead to a pattern where measures one occasion apart would be correlated p, measures two occasions apart would be connected as p^2, measures three occasions apart would be connected as p^3, and measures r occasions apart would be connected as p^r. This patterning was termed a *simplex model* by Guttman (1954). All structural restrictions work in this general way, and alternative restrictions are described later.

2.6. Graphic Model Expectations

In Figure 5.3 we list a few tracings through the example diagram that are consistent with the matrix algebra results. These algebraic results are often interpreted graphically by the "tracing rules of path analysis" (see Wright, 1921, 1984). For example, the second-order effect *ab* can be interpreted as a way to go backward from variable $Y(1)$ to variable $Y(3)$ in the diagram. The other nonzero elements of A^2 have the same two-arrow interpretation. Likewise, the third-order effect *abc* is seen in Figure 5.2(b) as the only way to use three arrows to go backward from any one variable to another within this diagram.

The graphic interpretation of the algebra can be summarized in three convenient graphic rules for the development of any expectation: (1) *Effects* are formed from the product of asymmetric coefficients (arrows). These effects can be traced in the diagram by following the specific number of arrows r ($r = 0$ to x) in each matrix of Table 5.3(a). (2) *Bridges* are formed as the product of zero or more asymmetric effects and one and only one symmetric coefficient (or sling). (3) *Connections* are formed from the addition of different bridges connecting the two end variables. These effects and components of covariance can be traced in

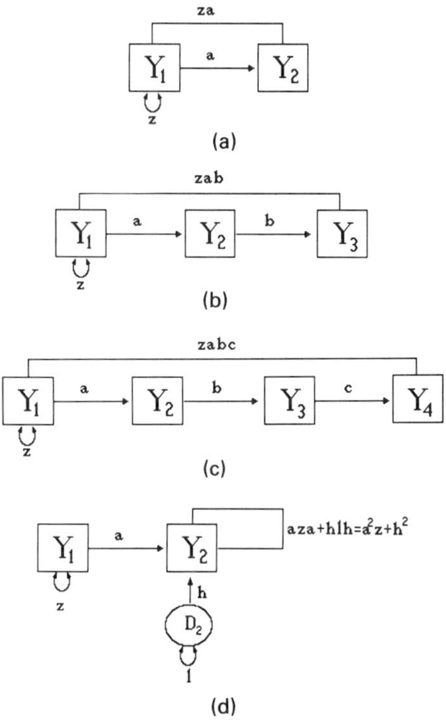

FIGURE 5.3. Selected graphic tracing of algebraic expectations: (a) tracing the covariance on one arrow; (b) tracing the covariance on two arrows; (c) tracing the covariance on three arrows; (d) tracing the variance.

the diagram, and these tracings follow each element of the connections of the equation in Table 5.3(b).

These three general tracing rules are only general interpretations of the algebra of Table 5.3. These three rules are slightly different than those of Sewell Wright (1921, 1934, 1984) because these rules work with more general association matrices (e.g., cross-products). As such, they are useful devices for the interpretation and demonstration of aspects of the model structure (for further details see McArdle and Horn, 1988). On the other hand, we do not want to overemphasize the importance of these graphic tracings because the computer programs make most of this tracing unnecessary (see next section).

One useful aspect of the graphic-algebraic approach comes when the algebra suggests alternative ways to draw the graph. In Table 5.3(d) we rewrite the same model in terms of a more *compact* set of matrices. In this representation we try to eliminate as many rows and columns of zeros as possible. This leaves a set of 4×4 matrices that produce the same set of connections for the manifest variables as the 7×7 matrices. This means the models are algebraically equivalent. The graphic display of this new model is presented as Figure 5.2(c), and it has some interesting properties. The exact graphic translation of the algebra demands the drawing of the four squares only. The arrows a, b, and c and the variance term z still retain their original position, but the disturbance variances are now placed as a two-headed arrow directly attached to each square. This provides an important algebraic interpretation to the graphic display of two-headed arrows—*the sling is always a partial connection.* Each sling reflects some residual connection of a variable after any arrows have been taken into account. This interpretation precisely matches the basic algebraic definition of S as a partial cross-product matrix (see appendix).

These algebraic devices allow several new graphic properties to be correctly used. This compact algebraic form also provides a convenient graphic device for the reduction of the size of the path diagram. The compact picture is equivalent because it produces the same connections, but it is smaller because it eliminates the drawing of residual variables. In this compact form it is somewhat hard to see that the implied residuals $uY(t)$ are uncorrelated with themselves and with the other variables. These assumptions are still in place, but it takes some practice to visualize these mathematical assumptions in the compact graphic.

As an alternative we can always draw a fully extended graphic where each variable has a residual, including $Y(1)$, and this picture is drawn as Figure 5.2(d). Here we have a balanced view of all variables because each has a residual $D(t)$. The effect from $D(1)$ to $Y(1)$ is equal to $w = \sqrt{z}$, so w is the standard deviation of $Y(1)$. $D(1)$ is the variable that gives the total variance to $Y(1)$. This model is useful because it allows the direct estimation of standard deviations, and this forces the variances to be nonnegative. This unusual picture is simply a graphic display of the 8×8 matrix representation that produces the same structure of the connections.

Figures 5.2(c), (d) are new graphic choices that reflect new algebraic choices. But there are no new translation rules used here—*the graphics are a precise reflection of the algebra.* We have found that these new devices can make researchers uncomfortable, and they may continue to use diagrams like Figure 5.2(b). This is a reasonable choice, although it makes the diagrams unnecessarily large. Other researchers may choose to rely on the traditional path diagram of Figure 5.2(a), and we think this is a poor choice. This traditional picture is incomplete, and this makes it difficult to accurately understand and convey the structual predictions of the model.

2.7. Model Estimation and Goodness-of-Fit

In general, our use of SEM algebra and graphics with current computer programming follows a simple plan: (1) The model of the scores defines the graph, (2) The graph defines the matrices: (3) The matrices define the nonlinear expectations about the connections. (4) The expectations can be compared to the observations. The practical aspects of the fitting of any change model requires more details of parameter identification and estimation. We deal with these technical issues in specific illustrations.

There are many computer programs which can be used to estimate the parameters of these SEMs. Aspects of these programs are described in the appendix. The best known is the LISREL program by Joreskog and Sorbom (1979, 1985). This is a landmark program, and we have used it extensively (Horn and McArdle, 1980; McArdle and Epstein, 1987). An example of this program applied to our sample problem is listed in the appendix. Another program is the COSAN program (see Fraser and McDonald, 1988). In related work (see McArdle, 1986b), we have developed a different programming language termed RAMIT to accomplish this same task using our own notation. These programs can be used to make an evaluation of structural hypotheses based on correlations, covariances, or means. The computer specification of a model using *fixed, free, or equal* parameters is relatively easy, and the estimation of *maximum likelihood estimates* (MLE) for free parameters is virtually automatic.

A few general principles of estimation are relatively important. First, the expected moments $E\{ij\}$ are derived for any set of parameters $P\{x\}$ and these are compared to the observed moments $O\{ij'\}$ from the data. This is usuallly done by setting up an estimation model where $R\{ij'\}$ is defined as the "residual" moment (i.e., $O\{ij'\} = X\{ij'\} + R\{ij'\}$). The estimation of parameters in such a model is virtually identical to a standard regression model except that: (1) the model expectations are nonlinear in the parameters; (2) the residuals are not scores, but they are average cross-products; and (3) the residuals need not be minimized with "least squares" or "minimum variance," but other functions can be chosen. In broadest terms, the computer algorithms try to find a set of parameters P that minimize the discrepancy between the expected and observed cross-products.

The nonlinear structural expectations provide both a way to estimate the model parameters and a way to test a structural hypothesis about the latent construct. We often define a discrepancy function f as a likelihood ratio statistic and obtain estimates for the parameters of the structural expectations by minimizing residuals according to the minimum chi-square criterion. These computer programs also provide details on standard errors, and on specific and overall measures of fit. (For further details, see the appendix.) From these we can form a *likelihood ratio criterion* (*LRC*) statistic from the scaled sums-of-squares squared differences between data observations and model expectations. As a result of this approach the statistical procedure also provides approximate MLEs, standard errors, and normal theory Z-tests.

The determination of an appropriate model for a particular set of data needs to be made on statistical, mathematical, and substantive grounds. It is easy to say "We want the simplest model that fits these data," but this is often hard to find. Essentially, the selection of a good model combines the problems of choosing a best regression model in regression analysis and in determining the best factor model in factor analysis. A variety of goodness-of-fit indices are available for model evaluation (see Cudeck and Browne, 1983; Sobel and Bohrnstedt, 1985; Matsueda and Bielby, 1986; Marsh, Balla, and McDonald, 1988). Most indices we use rely in some way on the size of the LRC. Put simply, if the likelihood of discrepancy is large enough, then the expectations do not match the observations, but if the likelihood of discrepancy is small enough then the expectations do match the observations. If we are interested in a subset of parameters, we may evaluate the *differences in the likelihood ratio* (*d*LRC) based on the *differences in the degrees-of-freedom* (*d*DF). Our analytic goal is to separate those models that clearly do not fit from those models that clearly do fit (for more details see Loehlin, 1987; McDonald, 1985; Cudeck and Browne, 1983; McArdle, 1986, 1988). To avoid being too dogmatic about uncertain issues, we simply try to demonstrate these model selection issues with the illustrations of the next section.

3. Results

In this section we use the previous SEM notation to examine some alternative structural models of change. These basic issues include change represented by (1) autoregressive models, (2) difference components models, (3) growth curve models, and (4) longitudinal factor models. The models we used here are not novel, and each has appeared in the SEM research literature. We use these empirical results to show how each kind of model provides a different view of developmental change for the same data set.

3.1. Autoregressive Models of Change

Figure 5.4 presents some alternative models of change based on the *autoregressive* or *Markov simplex* model. These change models are generally based on the linear equations listed in Table 5.4(a). In this model the mean deviations of variable $Y^*_{(t)}$ are predicted from the immediately preceding value $Y^*_{(t-1)}$. This model leads to a nonlinear structure of the covariances listed in Table 5.4(b). In this model each

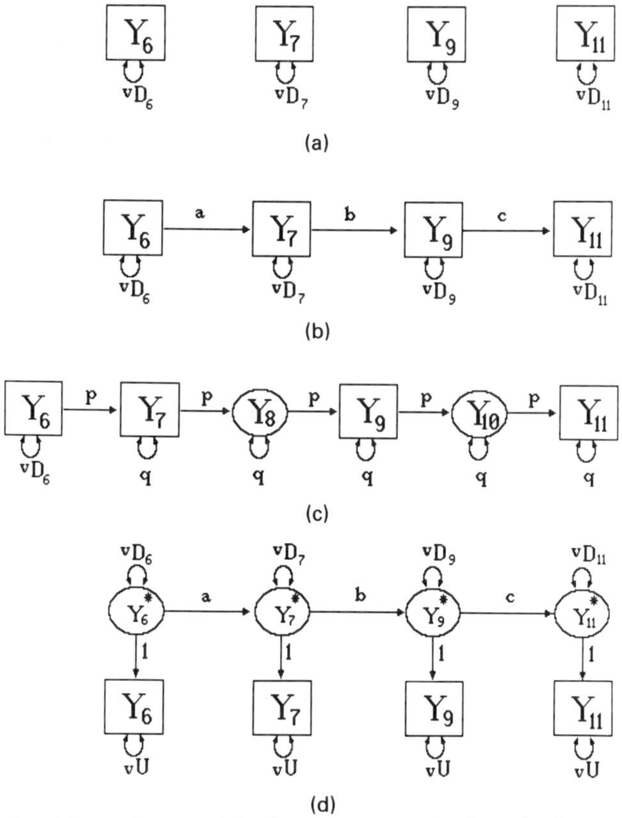

FIGURE 5.4. Alternative models for autoregressive-based change: (a) zero-covariance baseline; (b) Markov simplex process; (c) age-based Markov process; (d) reliable Markov process.

TABLE 5.4. Autoregressive Change Models Structure and Results

(a) The linear specification model

$$Y^*_{t,n} = a_1 Y^*_{t-1,n} + D_{t,n}, \quad \text{where}$$

$$Y^*_{t,n} = Y_{t,n} - \mathscr{E}\{Y_t\}, \quad \text{and}$$

$$\mathscr{E}\{D_t D'_t\} = vD_t,\ \mathscr{E}\{Y_{t-1} D_t t\} = 0,$$

where $Y_{t,n}$ = the raw score for individual n on occasion t

$Y^*_{t,n}$ = the mean deviate score for individual n on occasion t

$D_{t,n}$ = the disturbance variable

a_1 = the first-order regression coefficient

vD_t = the variance of the disturbance variable D_t

$\mathscr{E}\{x\}$ = the expected value of x

(b) The nonlinear structural expectations

$$C_{1,1} = \mathscr{E}\{Y_1 Y'_1\} = vD_{t1},$$

$$C_{2,2} = \mathscr{E}\{Y_2 Y'_2\} = vD_{t2},$$

$$C_{k,k} = \mathscr{E}\{Y_k Y'_k\} = vD_{tk},$$

$$C_{1,2} = \mathscr{E}\{Y_1 Y'_2\} = a_1 vD_{t1},$$

$$C_{1,k} = \mathscr{E}\{Y_1 Y'_k\} = a_1^k vD_{t1},$$

$$C_{2,k} = \mathscr{E}\{Y_2 Y'_k\} = a_1^{k-1} vD_{t2}.$$

(continued)

TABLE 5.4. Continued.

(c) Parameter estimates

Model parameter	#0 No covariance baseline	#1 Markov process	#2 Time-based Markov	#3 Age-based Markov	#4 Reliability Markov	#5 Age-based and reliability
Autoregression						
$Y_6^* \rightarrow Y_7^* := a_{7,6}$	0	.929*	1.009**	.996**	1.12*	1.09**
$Y_7^* \rightarrow Y_9^* := a_{9,7}$	0	.925*	1.009**	0	1.04*	0
$Y_9^* \rightarrow Y_{11}^* := a_{11,9}$	0	1.155*	1.009**	0	1.32*	0
$Y_7^* \rightarrow Y_8^* := a_{8,7}$	0	0	0	.996**	0	1.09**
$Y_8^* \rightarrow Y_9^* := a_{9,8}$	0	0	0	.996**	0	1.09**
$Y_9^* \rightarrow Y_{10}^* := a_{10,9}$	0	0	0	.996**	0	1.09**
$Y_{10}^* \rightarrow Y_{11}^* := a_{11,10}$	0	0	0	.996**	0	1.09**
Disturbance variances						
$Y_6^* \leftrightarrow Y_6^* := vD_6$	40.58*	40.58*	40.58*	40.58*	33.76*	33.10*
$Y_7^* \leftrightarrow Y_7^* := vD_7$	53.58*	18.49*	21.43**	13.84**	4.20*	2.40**
$Y_9^* \leftrightarrow Y_9^* := vD_9$	60.84*	17.39*	21.43**	13.84**	3.24*	2.40**
$Y_{11}^* \leftrightarrow Y_{11}^* := vD_{11}$	107.9*	26.83*	21.43**	13.84**	10.24*	2.40**
$Y_8^* \leftrightarrow Y_8^* := vD_8$	0	0	0	13.84**	0	2.40**
$Y_{10}^* \leftrightarrow Y_{10}^* := vD_{10}$	0	0	0	13.84**	0	2.40**
Measurement error variances						
$Y_6 \leftrightarrow Y_6 := vU_6$	0	0	0	0	6.81**	7.83**
$Y_7 \leftrightarrow Y_7 := vU_7$	0	0	0	0	6.81**	7.83**
$Y_9 \leftrightarrow Y_9 := vU_9$	0	0	0	0	6.81**	7.83**
$Y_{11} \leftrightarrow Y_{11} := vU_{11}$	0	0	0	0	6.81**	7.83**
$Y_8 \leftrightarrow Y_8 := vU_8$	0	0	0	0	0	0
$Y_{10} \leftrightarrow Y_{10} := vU_{10}$	0	0	0	0	0	0

Fixed parameters						
$Y_6^* \rightarrow Y_9^* := a_{9,6}$	0	0	0	0	0	0
$Y_7^* \rightarrow Y_{11}^* := a_{11,7}$	0	0	0	0	0	0
$Y_6^* \rightarrow Y_{11}^* := a_{11,6}$	0	0	0	0	0	0
$Y_T^* \rightarrow W_T^* := j_T$	1.0	1.0	1.0	1.0	1.0	1.0

(d) Goodness-of-Fit Indices

Model Index	#0 No covariance baseline	#1 Markov process	#2 Time-based Markov	#3 Age-based Markov	#4 Reliability Markov	#5 Age-based and reliability
Likelihood ratio	821	62	95	110	1	19
Degrees-of-freedom	6	3	7	7	2	6
Normal Z-test	22.	6.7	8.0	8.6	−.3	2.68
Measure of centrality	13%	86%	80%	77%	100%	97%
Likelihood index change	0%	93%	88%	87%	100%	98%
Interpretations	Some nonzero covariance	Markov better	Time-based not good	Age-based worse	Reliability better	Age-based possible

Notes: All parameters estimated with maximum likelihood function are designated by a single asterisk, and all equal parameters by a double asterisk.

covariance is a multiple of the pairwise autoregressive terms between each pair of scores. The autoregressive concepts have a long history in time-series analysis and now enjoy a widespread popularity in SEM (see Wold, 1949; Guttman, 1954; Joreskog, 1970; Cronbach and Furby, 1970; Hertzog and Nesselroade, 1987; Gollob and Reichardt, 1987).

This model is usually fitted to correlations or covariances and any pattern of change in the means *Mt* are ignored. The parameters a_t are the regression coefficients for the prediction of $Y_{(t)}$ from a prior time $Y_{(t-1)}$, so this reflects the *stability* component of the variance of Y over time. (These a_t parameters are labelled a, b, and c as in Table 5.2.) The implied disturbance terms $D_{(t)}$ have variance $vD_{(t)}$, and these represent the *independent and unstable* components of variation in this model. In this model the changes are reflected in the instability of the response between these times of testing.

Table 5.4(c) presents a numerical summary of the results of these kinds of models fitted to the WISC covariance matrix (of Table 5.1(c)). The second part of Table 5.4 gives a general way to list a variety of alternative models. In Table 5.4(c) we list the parameter estimates for the six separate models. In this section of the table each row is a model parameter defined with arrows, $\rightarrow$ and $\leftrightarrow$. Each column is a particular model and MLE or fixed values are listed within each column. In Table 5.4(d) we list several goodness-of-fit indices for each of the five models, and we add a brief interpretation of the results.

There are many models that we could fit to exploit the benefits of autoregression. The common theme of these change models is the specification and estimation of a variety of simple time-ordered models that separate stability from instability. The benefits of this model are enhanced when we combine autoregression with other variables in the well-known *cross lagged design* (see Duncan, 1969; Cook and Campbell, 1979; Kenny, 1979). In this example, we might start with the autoregressive structure for the reliable scores $W^*_{(t)}$ and then predict these scores from another variable, such as Mothers's Educational Level. This would help demonstrate how the currently independent changes in $Y_{(t)}$ are dependent upon some variables $X_{(t-1)}$ *besides* $Y_{(t-1)}$ itself. More complex autoregressive model extensions have potential for other organizations of change.

3.2. Autoregressive Examples

We begin most structural analysis by defining a baseline or starting-point model (see Sobel and Bohrnstedt. 1985). By writing the model labeled

#0, we assume that the entire covariance matrix can be fit using a model that only allows main diagonal disturbance variances for each variable. As Figure 5.4(a) shows, this is highly restrictive model requiring the off-diagonal covariances among all variables to be zero. The numerical results for the WISC are listed in the first column of Table 5.4(c).

Only four parameters are fit here, and these values are all variances terms (i.e., $6.37 \times 6.37 = 40.58$; $7.32 \times 7.32 = 53.58$, etc.). This initial model obtains an LRC = 82, with DF = 6, and the normal $Z = 22$ suggests this model is far from fitting the data. This model predicts zero covariances among all WISC scores, so we are not surprised by this relatively poor fit. In fact, this poor fit is a necessary starting point for any model: If this simple model fit the data, then we would have little reason to pursue more complex models. We primarily use these numerical results as a baseline against which to judge more complex models.

The next model, labeled #1, is the basic autoregressive model detailed in Table 5.2. This model is drawn as Figure 5.4(b), and, as the arrows show, this is a typical prediction model of scores over time. The numerical results for this model give autoregressive parameters of $a = .929$, $b = .925$, and $c = 1.155$. These are not values that make an optimal prediction of the scores. Instead, these values are the *most likely* given this hypothesized structure of the covariances. This model also estimates disturbance variances (i.e., $v_{D(t)} = 18.49$), so, by comparison to the baseline model, we can calculate the unexplained variance for each occasion (e.g., for occasion 2 we obtain unexplained variance as $18.49/53.58 = 34.5\%$). This model fits the data with LRC = 62 on DF = 3 as a test of the hypothesis of proportional constraint on three covariances. This simple model has a $Z = 6.7$, so, using either the conventional $Z = 1.96$ or $Z = 2.56$ cutoff levels, we would reject the idea that the residuals are collectively zero. However, we do not completely reject the hypothesis represented by this model because it yields a 93% improvement in the overall fit.

Model #2 is a more restrictive version of the Markov process model. Here we assume there is only one autoregressive coefficient p and one disturbance term q. This is accomplished by forcing all autoregressive parameters a, b, and c to be equal to a single parameter p. Likewise, the parameters h, j, and k are forced to be equal to a single parameter q. This model is a time-series based model that assumes that changes over any one interval of time are the same as changes over any other interval of time. These restrictions lead to a structure that has strict proportionality of covariances based entirely on the number of occasions of measurement. This model also forces the variance at each time point to be equal (for $T = 2, 3, 4$). This collective assumption of *stationarity* is a typical requirement of most causal analyses, and in this model it is tested empirically.

This retest-based Markov process model fits the data with LRC = 95 on DF = 3 for $Z = 8$ when fitted to the covariance matrix. This model is an 88% improvement over the baseline model #0, so it seems useful. The specific change due to these four additonal restrictions can be obtained by comparing #2 fit to the fit of Markov model #1. The difference between models #1 and #2 shows the time-based restrictions #2 produce a change of dLRC = 33 on dDF = 4 and this is a loss of fit of 47%. These equality constraints are probably not acceptable fitted to these covariances and the changes between testings are not equal.

Model #3 is diagrammed in Figure 5.4(c) and uses latent variables to structure a different Markov model. In these longitudinal data the distances between the occasions of measurement are not equal. These children were measured at ages 6, 7, 9, and 11, so the difference between occasions 1 and 2 is one year, but the other occasions are two years apart. Thus, model #2 presumes the changes are equal over different intervals of time. Model #2 might be an acceptable model, especially if the occasions of measurement were chosen to represent equal developmental intervals. But more typically, the choice of occasions of measurement reflects a choice of convenience, and we need to restructure the mathematical model to account for these choices.

Model #3 tries to take this age distance into account. In this model we have added latent variables for the missing occasions of measurement at ages 8 and 10. The model is identified by forcing equality constraints on all autoregressive a_t terms and on the four disturbance terms D_t over time. In this model the connection between ages 6 and 7 is p, but the connection between ages 7 and 9 is p^2, and the connection between ages 9 and 11 is p^2. The underlying structure of the other connections is also patterned according to these proportions. For example, the connection between ages 6 and 11 is structured as $p \cdot p^2 \cdot p^2 = p^5$. The latent variables at age 8 and age 10 are simply used to provide a placeholder for the timing pattern of the autoregressive parameter. Most critically, in this model the single parameter p is interpretable as the stability coefficient for any one-year interval of time. The addition of more precise nonlinear constraints are possible (using extra scalar model constraints in programs like COSAN).

The maximum likelihood estimates of autoregression $p = .996$ and disturbance $q = 3.72$ are used to make a rigid prediction about the age-based structure of these data. The goodness-of-fit of this age-based model is indexed by the LRC = 110 on DF = 7. This model is slightly worse than the previous retest-based model, and this is disappointing. After all, the single parameters p and q can be used to describe a simple system of independent changes over any one-year interval.

This age-based model #3 is usually rejected by developmental theory. In this model we assume each year between 6 and 11 reflects the same amount of developmental changes in cognition. It seems reasonable to assume that more changes might occur between, say, ages 6 to 7 than between ages 10 to 11. As it now stands, this model provides a very simple structure to the data, but this is probably not true of cognitive development measured by the WISC. In a substantive investigation we would probably pursue the line of reasoning in more depth. However, since we are now most interested in alternative methodologies, we turn to a different kind of latent variable model.

The next model of Figure 5.4(d) includes both the autoregressive and an unreliability component v_U. This model adds one latent variable Y_t^* at each time point to separate the reliable variance from the unreliable variance. We further assume that all covariance in the data can be accounted for by the autoregressive structure among these reliable variables. For the purposes of identification we have fixed the scale of the latent variable Y_t^* by forcing the loading to the unit-valued, and we have estimated only one unique variance term for the entire series. Other forms of this reliability model are possible (e.g., the "quasi-Markov" model of Joreskog, 1970), but this seems like a sensible model in the presence of large errors of measurement in the repeated observations.

This reliability model allows a more relaxed structure to the data because we allow all variances to have an extra additive component. The resultant fit is much better, with LRC = 1 on DF = 2. Because the coefficients of autoregression are related to purely reliable variables, they are necessarily larger (see Goldberger, 1973). In this case we obtain $a = 1.12$, $b = 1.04$, and $c = 1.32$ (compare #1 and #4), but we also obtain larger standard errors for these coefficients. This is our first use of a latent variable model representing a measurement system, and we return to this model in several examples to follow.

Model #5 is a mixture of the age-based autoregression constraint model #3 with the reliability model #4. This model tests the same age-based constraints, but, since we have added the latent variable, it fits this restriction only to the covariances. This model fits with LRC = 19 on

DF = 6, and this is a loss in fit of dLRC = 18 on dDF = 4 so this idea may remain unacceptable. On the other hand, the discrepancy dLRC = 18 between latent variable models #4 and #5 is much lower than the discrepancy dLRC = 48 between manifest variables models #1 and #3. Since these are the same constraints for the same dDF = 4, this result demonstrates a loss of statistical power to detect the age-based restrictions. This loss of power is a cost of the estimation of the reliability coefficient v_U. These benefits and problems of power analysis are often overlooked in the testing the hypotheses on reliable variables (see Matsueda and Bielby, 1986).

3.3. *Difference Component Models of Change*

Figure 5.5 presents some alternative models of change based upon concepts of *difference components* or the *Weiner simplex.* These change

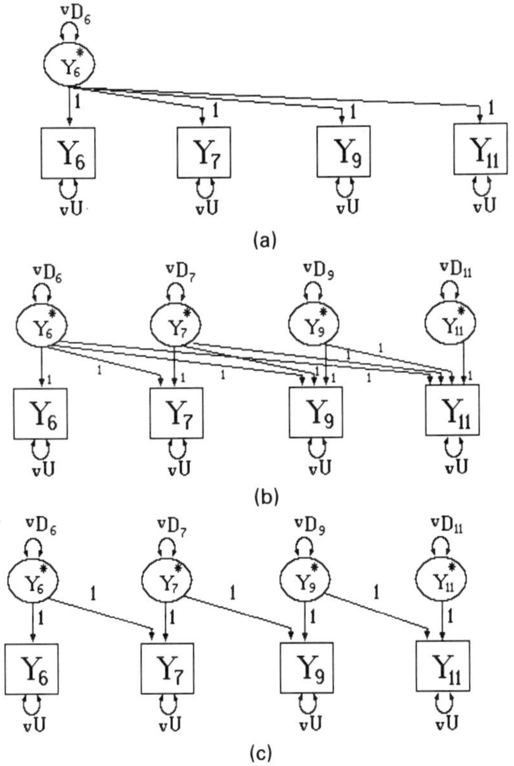

FIGURE 5.5. Alternative models for difference-based change: (a) equal variance-covariance baseline; (b) Wiener simplex process; (c) moving-average difference.

TABLE 5.5. Difference Components—Based Change Models and Results

(a) Linear model specification

$$Y^*_{t,n} = D_{1,n} + D_{2,n} + D_{3,n} + \cdots + D_{t,n} + U_{t,n}, \quad \text{so}$$

$$D_{t,n} = Y^*_{t,n} - Y^*_{t-1,n}, \quad \text{and where}$$

$$Y^*_{t,n} = Y_{t,n} - \mathscr{E}\{Y_t\}, \quad \text{and}$$

$$\mathscr{E}\{D_t D'_t\} = vD_t, \qquad \mathscr{E}\{D_t D'_{t-k}\} = 0,$$

$$\mathscr{E}\{U_t U'_r\} = vD,$$

where $D_{t,n}$ = a difference component for individual n at time t
vD_t = the variance of the difference component

(b) Nonlinear structural model

$$C_{1,1} = \mathscr{E}\{Y_1 Y'_1\} = vD_1 + vU,$$

$$C_{2,2} = \mathscr{E}\{Y_2 Y'_2\} = vD_1 + vD_2 + vU,$$

$$C_{k,k} = \mathscr{E}\{Y_k Y'_k\} = vD_1 + vD_2 + vD_3 + \cdots + vD_k + vU,$$

$$C_{1,2} = \mathscr{E}\{Y_1 Y'_2\} = vD_1,$$

$$C_{1,k} = \mathscr{E}\{Y_1 Y'_k\} = vD_1,$$

$$C_{2,k} = \mathscr{E}\{Y_1 Y'_3\} = vD_1 + vD_2.$$

(continued)

TABLE 5.5. Continued.

(c) Parameter estimates

Model parameter	#0 Equal covar. baseline	#1 Weiner process	#2 Equal differences	#3 Moving average	#4 Reliability weiner	#5 Reliability and equal
Difference components variance						
$D_6 \leftrightarrow D_6 := vD_6$	51.70*	41.34*	40.58*	40.58*	37.45*	37.09*
$D_7 \leftrightarrow D_7 := vD_7$	0	18.75*	21.44**	18.75*	8.94*	11.29**
$D_9 \leftrightarrow D_9 := vD_9$	0	17.39*	21.44**	62.57*	7.08*	11.29**
$D_{11} \leftrightarrow D_{11} := vD_{11}$	0	28.30*	21.44**	55.06*	18.25*	11.29**
$D_6 \rightarrow Y_6^*$	1.0	1.0	1.0	1.0	1.0	1.0
$D_6 \rightarrow Y_7^*$	1.0	1.0	1.0	1.0	1.0	1.0
$D_6 \rightarrow Y_9^*$	1.0	1.0	1.0	0	1.0	1.0
$D_6 \rightarrow Y_{11}^*$	1.0	1.0	1.0	0	1.0	1.0
$D_7 \rightarrow Y_7^*$	0	1.0	1.0	1.0	1.0	1.0
$D_7 \rightarrow Y_9^*$	0	1.0	1.0	1.0	1.0	1.0
$D_7 \rightarrow Y_{11}^*$	0	1.0	1.0	0	1.0	1.0
$D_9 \rightarrow Y_9^*$	0	1.0	1.0	1.0	1.0	1.0
$D_9 \rightarrow Y_{11}^*$	0	1.0	1.0	1.0	1.0	1.0
$D_{11} \rightarrow Y_{11}^*$	0	1.0	1.0	1.0	1.0	1.0
$Y_T^* \rightarrow W_T^*$	1.0	1.0	1.0	1.0	1.0	1.0
Measurement error						
$Y_6^* \leftrightarrow Y_6^* := vU_6$	13.99**	0	0	0	4.62**	4.84**
$Y_7^* \leftrightarrow Y_7^* := vU_7$	13.99**	0	0	0	4.62**	4.84**
$Y_9^* \leftrightarrow Y_9^* := vU_9$	13.99**	0	0	0	4.62**	4.84**
$Y_{11}^* \leftrightarrow Y_{11}^* := vU_{11}$	13.99**	0	0	0	4.62**	4.84**
$Y_8^* \leftrightarrow Y_8^* := vU_8$	0	0	0	0	0	0
$Y_{10}^* \leftrightarrow Y_{10}^* := vU_{10}$	0	0	0	0	0	0

(d) Goodness-of-fit indices

Model Index	#0 Equal covar. baseline	#1 Weiner process	#2 Equal differences	#3 Moving average	#4 Reliability weiner	#5 Reliability and equal
Likelihood ratio	179	81	95	477	41	62
Degrees-of-freedom	8	6	8	6	5	7
Normal Z-test	11.	7.4	7.9	17.	5.0	6.2
Measure of centrality	65%	83%	81%	31%	91%	87%
Likelihood index change	0%	56%	47%	167%	77%	65%
Interpretations	Some nonequal covariances	Weiner better	Time-based not bad	Simple MA_1 bad	Reliability better	Equal not-good

models are generally based around the linear equations in Table 5.5(a). In these models D_{t-1} is an unobserved difference component with variance vD_{t-1}. This component supposedly emerges at time t and persists in the mean deviations of variable Y_t^*. This model leads to a nonlinear structure of the covariances of Table 5.5(b), where each covariance is a sum of the number of common or overlapping variance components vD_t. In this model we presume the observations Y_t reflect the buildup or decay of several additive components D_t. The impact of each component, and the scale of the developmental process, is given by the variance term vD_t. These structural restrictions are a result of our model, which assume that *changes are independent, additive, and persistent.* This alternative model reflects a simple developmental process but does not use score prediction over time. The difference components concepts have a long history in data analysis, but these models are not widely used in SEM.

Perhaps the most important aspect of this treatment is displayed in the figures themselves. These simplex models of Figure 5.4 and Figure 5.5 look very different from a developmental point of view. Brilliant treatments of these difference component models have been presented by Jones (1962) and by Joreskog (1970). In particular, Joreskog distinguished the autoregressive simplex as a "Markov process" and the component model as a "Weiner process," and said,

> The scale for the Markov simplex is just a logarithmic transformation of the scale for the Weiner simplex. In the Markov simplex, correlations correspond to distances between scale points, whereas in the Weiner simplex, correlations correspond to square roots of ratios between scale points. . . . In contrast to the Markov simplex, the Wiener simplex is scale dependent (pp. 128–129).[1]

The Markov simplex is now very popular in SEM but the Weiner simplex can often be more useful in repeated measures analyses (also see Short, Horn, and McArdle, 1983).

3.4. Difference Component Examples

Table 5.5(b), (c) present a numerical summary of the results of difference components models fitted to the WISC covariance matrix of Table 5.1(c). We begin these structual analyses by defining another baseline model labeled #0. In this case we assume one unique component uY_t for each

[1] Reprinted with permission from the *British Journal of Mathematical and Statistical Psychology,* Joreskog, K. G. "Estimation and testing of simplex models," **23** (1970).

variable Y_t and one common component of correlation D_6 with variance vD_6. This model makes a structural prediction where all variances are equal and all covariances are equal. This model is formally equivalent to a single-factor model with equal loadings and equal uniqueness, and it can be fitted this way. If we are primarily interested in different changes over time, we would hope this model would not fit the data very well. The parameters of $sD_6 = 7.19$ and $s_U = 3.74$ define the ratio of common and unique variance (and this is a tau-equivalent form of coefficient alpha.) The fit of LRC = 179 on DF = 8 is probably not acceptable, and we use this as our baseline result.

Model #1 is a Weiner simplex model described earlier. In this model we assume a first component D_6 that starts at age 6, a second component D_7 that starts at age 7, a third component D_9 that starts at age 9, and a fourth component D_{11} that starts at age 11. The parameter estimates give the impact of each component as variances equal to 41.34, 18.75, 17.39, and 28.30. The first component D_6 is a score for all of the changes that have occurred up to age 6, but the other three components reflect changes within the specified interval of time. Thus, up to age 6 we find an individual variation in scores indexed by 41.34, between ages 6 and 7 we have changes in scores indexed by 18.75, between ages 7 and 9 we have changes in scores indexed by 17.39, between ages 9 and 11 we have changes in scores indexed by 28.30. The model does not require the age interval to be equal because it is simply describing the additive changes in variability and covariability that occur over this time interval.

The restrictive structure of the Weiner process model is different from the previous Markov process model. In the Weiner model we use these four variance terms to structure the covariances *equally down the columns*. In this model, for example, the variance of variable 1 is supposedly equivalent to the covariances of variables 1 and 2, and 1 and 3, and 1 and 4. Recall that in the Markov model the covariances were assumed to be proportional across diagonals of the matrix (see Table 5.2). The test statistic of LRC = 81 on DF = 6 is non trivial, and our improvement in fit over the equal covariance baseline is 56%.

There are many ways to expand these concepts, and the model labeled #2 takes this logic in one direction. This model forces equality constraints across the variances of the three difference components. In general, this is a test of the hypothesis that an equal amount of development occurs during the three time intervals between measurement occasions. However, in this example the age between testings is not equal. So, since the actual time between ages 6 and 7 is half as long as the time between ages 7 and 9 and 9 and 11, this model actually tests the idea that twice as much development occurs during the one-year interval between ages 6 and 7 as any other comparable age period.

Model #2 fitted to these data obtains a value of $vD_t = 21.44$ for the equal changes, the LRC = 95 on DF = 8, and this fit is 47% better than our equal-covariance baseline model #0. A formal test of the equality of the first differences here can be calculated by the difference between models #2 and #1: dLRC = 14 on dDF = 2 with a percentage loss of fit of 17%. This difference is small but nontrivial, so the hypothesis of twice as much development between ages 6 and 7 is not really bad but is also not really critical.

The model labeled #3 and drawn in Figure 5.5(c) uses these difference components in another way. Here we organize the components model so that only the pairwise Y_t and Y_{t+1} variables share a component D_t. The first component is a starting point D_6, but the rest of the components represent a *moving average or random shocks* model from time series analysis (after Wold, 1954). These models are highly restrictive because they require a zero covariance among all pairs of scores more that one occasion apart (i.e., $C_{i,j} = 0$ where $j > i + 1$). The fit of this model to these WISC data obtains an LRC = 477 on DF = 6 and, with a −167% loss of fit, this model is worse than our equal covariance baseline #0. Nevertheless, this is a popular model that matches some developmental theory, can be extended in several ways (e.g., MA_2), and can be fitted using SEM.

The model labeled #4 improves the model in another direction. This model could be drawn by taking Figure 5.5(b) and adding four latent variables to include an estimation of unreliability as a model parameter. This model obtains an LRC = 41 on DF = 5, a 77% improvement in fit over the baseline. This fitting of the Weiner simplex to the latent variables uses one more parameter v_U than the manifest variable Weiner simple model #1. The difference in fit between models #1 and #4 is indexed by the dLRC = 40 on dDF = 1, a 49% improvement.

The final model labeled #5 is a direct combination of the equal differences model #2 and the reliability model #4. As before, we test hypotheses about the equality of the first differences in the context of reliable variables. The goodness of fit obtained is LRC = 62 on DF = 7 and this is a change of dLRC = 21 on dDF = 2 compared with the unequal but reliable #4. This difference is slightly greater than the same difference between the manifest variable models #1 and #2 where dLRC = 14 on dDF = 2. This is an unusual empirical result—manifest variable hypotheses of this kind are usually more powerful than reliability corrected hypotheses—but it come as a result of the equal reliability assumption. In any case, this is just another illustration of the possibilities of testing hypotheses with latent variables.

3.5. Latent Growth Models of Change

Figure 5.6 presents a different model of change based upon concepts derived from *growth curve analysis*. These change models are generally based around the linear equations Table 5.6(a). In this model the cores $Y_{(t)}$ are a reflection of some individual difference components termed level L and slope S. The latent slope S variable is related to the observed scores with coefficients b_t, so these coefficients are termed the *basis coefficients*. We also add an unobserved error component E, which has zero mean, equal variance vE at each time, and these E scores are uncorrelated across time. In this model we presume the observations $Y_{t,n}$ reflect the buildup or decay of several additive components C_n. The impact of each component, and the scale of the developmental process, is given by the variance terms vC and the basis of shape b_t. This model reflects a simple developmental process that does not emphasize score prediction over time. This model is termed a growth curve because the basic parameters describe a systematic pattern of individual difference in change over time. These growth curve concepts have a long history in longitudinal data analysis (see Tucker, 1958, 1966; Rao, 1958; Hultsch, Nesselroade, and Plemons, 1976), but they are not yet widely used in SEM (see Meredith and Tisak, 1984, this volume; McArdle and Epstein, 1987; McArdle, 1986, 1988).

This model leads to structural predictions (Table 5.6(b)) where the means, standard deviations, and correlations are all expected to be

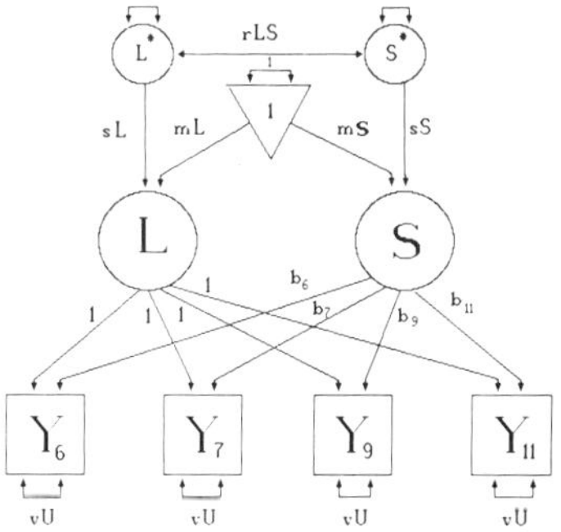

FIGURE 5.6. Generic path diagram for latent growth curve change models.

TABLE 5.6. Growth Curve-Based Change Models and Results

(a) Linear model specification

$$Y_{t,n} = L_n + b_t S_n + U_{t,n}, \quad \text{and}$$

$$\mathscr{E}\{LL'\} = vL, \quad \mathscr{E}\{SS'\} = vS, \quad \mathscr{E}\{LS'\} = cLS, \quad \mathscr{E}\{U_t U_t'\} = vU_t, \quad \text{and}$$

$$\mathscr{E}\{L\} = mL, \quad \mathscr{E}\{S\} = mS,$$

where $Y_{t,n}$ = the raw scores for individual n at time point t

L_n = the "level" scores at all times

S_n = the "shape" scores at all times

U_t = the "unique" or disturbances scores at time t

b_t = the "basis" coefficients at time t

vX = the variance of the Xth component

cSL = the covariance of levels and shapes

mX = the mean of component X

Also sX = the standard deviation of component X

rXY = the correlation of component x and Y

(b) Nonlinear structural model of cross-products

$$\mathscr{E}\{vY_t\} = vL + b_t vS\, b_t' + 2b_t cSL + vU_t,$$

$$\mathscr{E}\{cY_{t,t+i}\} = vL + b_t vS\, b_{t+i}' + b_t cSL + cSLb_{t-i}', \quad \text{and}$$

$$\mathscr{E}\{mY_t\} = mL + mS\, b_t.$$

(c) Parameter estimates

Model parameter	#0 Level only baseline	#1 Linear basis	#2 Retest basis	#3 Latent basis	#4 Absolute scaling	#5 Linear and errors
Factor loadings						
$S^* \rightarrow Y_6 := b_6$	.0	.000	.000	.000	.399*	.000
$S^* \rightarrow Y_7 := b_7$	.0	.186	.333	.272*	.559*	.186
$S^* \rightarrow Y_9 := b_9$	.0	.579	.667	.598*	.756*	.579
$S^* \rightarrow Y_{11} := b_{11}$	.0	1.000	1.000	1.000	1.000	1.000
$L^* \rightarrow Y_6$	1.0	1.0	1.0	1.0	0	1.0
$L^* \rightarrow Y_7$	1.0	1.0	1.0	1.0	0	1.0
$L^* \rightarrow Y_9$	1.0	1.0	1.0	1.0	0	1.0
$L^* \rightarrow Y_{11}$	1.0	1.0	1.0	1.0	0	1.0
Factor standard deviations						
$L^{**} \rightarrow L^* := sL$	3.67*	5.64*	5.41*	5.61*	.0	5.52*
$S^{**} \rightarrow S^* := sS$	.0	4.76*	4.77*	5.22*	10.3*	3.74*
$U_6^{**} \rightarrow U_6^* := sU_6$	12.9**	3.27**	3.28**	2.98**	3.43**	3.49*
$U_7^{**} \rightarrow U_7^* := sU_7$	12.9**	3.27**	3.28**	2.98**	3.43**	3.58*
$U_9^{**} \rightarrow U_9^* := sU_9$	12.9**	3.27**	3.28**	2.98**	3.43**	2.38*
$U_{11}^{**} \rightarrow U_{11}^* := sU_{11}$	12.9**	3.27**	3.28**	2.98**	3.43**	4.29*
Factor correlations						
$S^{**} \leftrightarrow L^{**} := rSL$	0	.678*	.672*	.561*	0	.951*
Factor means						
$1 \rightarrow L^* := mL$	31.4*	19.2*	17.2*	18.1*	.0	19.8*
$1 \rightarrow S^* := mS$	.0	27.7*	28.5*	28.6*	46.5*	27.8*

(continued)

TABLE 5.6. Continued.

(d) Goodness-of-fit indices

Model Index	#0 Level only baseline	#1 Linear basis	#2 Retest basis	#3 Latent basis	#4 Absolute scaling	#5 Linear and errors
Likelihood ratio	1687	92	90	16	84	77
Degrees-of-freedom	11	8	8	6	8	5
Normal Z-test	31.	7.7	7.6	2.2	7.3	7.2
Measure of centrality	2%	81%	82%	98%	83%	84%
Likelihood index change	0%	95%	95%	99%	95%	96%
Interpretation	Some growth	Linear better	Retest as good	Latent good	Intercept needed	Unequal errors

proportional. This kind of model prediction implies the growth curve model must be fitted to the cross-products matrix (Table 5.1(d)) or scaled (Table 5.1(e)). This use of cross-products is often confusing, but it is necessary for any model where the means and the covariances share joint restrictions. The unobserved components C in this model have variances vC, correlation rC, and means mC, and these explain the variances vY, correlations rY, and means mY among the observations Y_t. The basis parameters b_t are critical in this model because they are used in the predictions of the means, variances, and correlations. The basis coefficients b_t are the structural constants of proportionality that run through all statistical expectations.

The introduction of the means is an important feature of this model. In a cross-products matrix the means are considered as a summary statistic representing the average cross-product between any variable Y_t and the unit constant 1. We add this constant as an observed variable within the data set. This variable, labelled "1" or "UNIT," is observed but not measured so we draw it in our diagrams as a triangle. Now, the graphic rules are easy to state: (1) Any arrow from the triangle to another variable Y is a location parameter or mean. (2) If there are other arrows into variable Y, the mean parameter is an intercept in this equation. (3) We should not point any arrow towards the triangle. Means can be added to any model in this notation without any change in the path analysis tracing rules.

This generic growth curve model allows one component of variance for the intercept or level and one for the linear slope. These are the scores for the parameters of each individual's linear curve, but, of course, the individual curves might not be linear. In this case the collection of curves may not be well characterized by two components of intercept and slope parameters, and we may need to account for changes in curvature by fitting more and more latent components. Another option we pursue here is identical to the linear growth model except: we *estimate the basis b_t coefficients from the summary statistics.* This model is termed a *latent growth model* (*LGM*), and we give an example of this SEM in the next section.

The path diagrams of Figure 5.6 seem much different than those of the previous simplex models. The general move toward common latent components nicely matches the difference score simplex of Figure 5.5, but it does not match the Markov simplex models of Figure 5.4. In broad terms the LGM is a SEM that uses age as a dimension or an axis to evaluate change. In this sense the LGM is *timeless,* and this model nicely matches the principles enunciated by many others who have discussed a developmental function (e.g., Wohlwill, 1970, 1973, 1988; Horn, 1972). It is also historically relevant that this LGM resembles the rate of growth as a latent variable used by Sewell Wright in the original path diagrams (as in Figure 5.1). More critically now, this LGM model produces testable hypotheses about the variances, covariances, and means together.

3.6. *Latent Growth Model Examples*

Table 5.6(c), (d) present a numerical summary of the results of some growth curve models fitted to the WISC cross-product matrix of Table 5.1(e). We begin this structural analysis by defining a baseline model labeled #0, where we assume one level variable L, one error variable E, but no slope S. The latent level L variable is related to all observed scores Yt in the same way, and similar to the starting point D_6 in the previous difference models (as in Table 5.5(a)). This model requires the repeated-measures variances to be equal, the covariances to be equal, and the means to be equal. The model obtains parameter estimates of $sL = 3.67$ (so $vL = 13.4$), $dE = 12.9$ (so $vE = 166.4$), $mL = 31.4$, and a fit of $LRC = 1687$ on DF = 11. These restrictions do not fit these data very well.

The next model fitted is labeled #1, and it adds a slope variable S that has a linear basis. This is a SEM-based analysis of the *linear polynomial model.* In this model the b_t relations between S and Y_t are calculated as $b_t = [a_t - a_T]/[a_1 - a_T]$, where a_t is the actual group age at testing $t = 1$ to T. This means b_t reflects the interval of time between the ages as a ratio of the overall time between the first and last measurements. In this example, the $b_1 = .000$ reflects no time passed by age 6, the $b_2 = .186$ means 18·6% of the overall time has passed by age 7, the $b_3 = .579$ means 57.9% of the overall time has passed by age 9, the $b_2 = 1.000$ means 100% of the overall time has passed by age 11 (the more precise group ages of Table 5.1 were used in these calculations). Since the basis is forced to be linear with age, we can interpret the S scores as latent slopes.

The parameters of this familiar linear model are directly interpretable in the usual way. The mean level $mL = 19.2$ is the starting point or the mean at age 6, and the mean slope $mS = 27.7$ is the change in the mean level between ages 6 and 11. These means mL and mS are the usual parameters obtained from a MANOVA model of the same structural form. The positive increase in mean mS makes a prediction about a positive increase in the observed means because $mY_t = [b_t mS + mL]$. The variance around these scores is indexed by $sL = 5.64$ ($vL = 31.8$), which is the variance at age 6, $dS = 4.76$ ($vS = 22.7$), which indexes the variation in the way individuals change, and $dE = 3.27$ ($vE = 10.7$), which is the additional variance specific to each occasion. The correlation among the level and slope scores is $rSL = .678$. The substantial variance of slopes vS makes a prediction about a positive increase in the observed variances because $dY_t = [b_t sS + sL + 2\, sL\, sS\, rSL]$. Also the covariance between each occasion is based on a similar function of the b_t coefficients and the component variances and covariances (see Table 5.5(b)). The error variance can be used to index the unreliability if the error variance is compared to the overall score variance within each tme point (for details, see McArdle and Epstein, 1987). This linear model #1 fits these data with LRC = 92 on DF = 8. Although this model does not fit the data perfectly ($Z = 7.7$), linear change is a dramatic 95% improvement over our prior no-change baseline.

The model labeled #2 fits the same model with a different basis—an *equal interval* or *retest* basis. This basis reflects an equal interval scaling of the differences between the four occasions of measurement. That is, there are three intervals of time between the occasions, so we scale each of these as 1/3 apart; that is $b_t = \{0, 1/3, 2/3, 1\}$. This model assumes that the developmental process is indexed by an equal interval process that is linear with time of measurement but is not linear with age. This alternative basis is fitted, and we obtain LRC = 90 on DF = 8. This equal interval or retest model is certainly as good as the linear age model, and it may even be better. Still the overall model is not a perfect fit ($Z = 7.6$), so we examine other solutions as well.

There are many ways to proceed at this point. In the typical ANOVA or polynomial approach we add another common component termed a *quadratic score* Q with a fixed basis b_t^2. This model requires four additional parameters: mean mQ, variance vQ, and correlations rQL and rQS. Instead, we now fit a model labelled #3 and termed a *latent growth model* (LGM). This model is identical to the linear growth model, except we *estimate the basis coefficients from the summary statistics*. This model is identical to a confirmatory two-factor model where (1) one of the factors (L) is fixed and used as a reference point for the other (S), (2) the model is fitted to both the means and the covariances, and (3) in this case there is only one uniqueness variance dE_t for all measures. This model is simpler than the quadratic polynomial model because it adds only two new parameters, b_2 and b_3.

The estimated b_t can be labeled a *latent time scale* or *metameter,* which can be compared to the actual age curve. The results from this model gives the parameter $b_1 = .272$, and this is 46% larger than $b_1 = .186$ based on the actual age differences. This means the latent developmental process is 46% more rapid during ages 6 to 7 than at other times between ages 6 and 11. The second latent parameter, $b_2 = .598$, is very close to the manifest $b_1 = .579$, so perhaps the latent developmental process aligns itself with manifest age by age 9. The variance of the shape, $sS = 5.22$, is about 9% larger than in the linear model #1, so more individual differences in shape are captured by this model. This LGM fits the data with LRC = 16 on DF = 6, and this is a clear improvement over the linear models with dLRC = 76 on dDF = 2. Indeed, it seems likely that the overall developmental function changes in ways that are monotonic with age but are not linear with age.

This factor analytic interpretation leads to the obvious possibility of fitting a simpler one factor model with just S. The single dimension of slope here describes the empirical curve without any intercept, so we use the general term *curve* to refer to a *connected line.* Since this curve is latent, b_t is allowed to change with age a_t and we make no presumption about any ratio or absolute zero point. In model #4 we estimate $b_t = 1.399$, .559, .756, and 1.00], and these represent the ratio of changes on this function. The difficulty here comes because these factor scores require a broader interpretation of the function S than in the previous model. Now, for example, low factor scores could come from patterns of curves that are completely different at all points, or from patterns of curves that are identical in form but are lower in size. Thus, what may be separate concepts of level and shape are confounded in this model. In this one-factor model the single-factor score is a curve, and we use the concept of *similarity* to describe the factor scores S_n. This model of a single dimension of change is very restrictive and LRC = 84 on DF = 8. The loss of the intercept component L resulted in a loss of dLRC = 68 on dDF = 2 in fit. In this case the lack of an intercept does not seem like a reasonable idea. Other aspects of this model are described by McArdle and Epstein (1987) and McArdle (1988).

The final model, #5, tries to improve the fit of the age-based linear model #1 by allowing for unreliability in the manifest variables. Errors of measurement are not separable from errors of the function within this mathematical model. In this model we can allow the unique variance of error term E_t to be different at each time point and this forces all restrictions on all covariances but has no restrictions on the variances. This linear model is fitted and we obtain LRC = 77 on DF = 5. The addition of unequal error variances improves the model by dLRC = 15 in dDF = 3 (compared with #1). So, while there may be different influences of the error terms within each age, the linear model still does not fit the means and covariances very well.

3.7. *Longitudinal Factor Models of Change*

Figure 5.7 presents a generic model of change based upon concepts from the *longitudinal factor analysis* model. These structural change models are generally based around the linear equations listed in Table 5.7(a). These models are substantially different from all previous models because they require multiple variables measured with each occasion (so they cannot be fitted to the data of Table 5.1). Within each occasion of measurement the mean deviations of several variables Y_t^* are predicted from some common factor scores P_t^* with factor loading j_t and some unique factor scores U_t^*. Between each occasion the relationships are largely patterned between these common factor scores P_t and P_{t+x}. This model leads to a nonlinear structure of the covariances listed in Table 5.7(b). In these expressions each covariance is a multiple of the factor loadings and factor covariances within and between occasions. These longitudinal models have a long history in factor analysis and are popular in SEM (see Joreskog, 1979; Hertzog and Nesselroade, 1987;

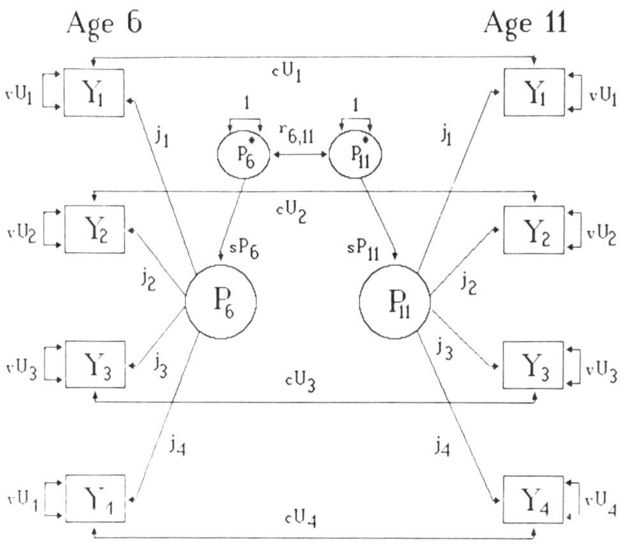

FIGURE 5.7. Generic path diagram for longitudinal factor change models.

TABLE 5.7. Longitudinal Factor Analysis–Based Change Models and Results

(a) Linear model specification

$$Y_{m,t,n} = j_{m,t}P_{t,n} + U_{m,t,n}, \quad \text{and}$$
$$Y_{m,t+1,n} = j_{m,t+1}P_{t+1,n} + U_{m,t+1,n},$$
$$\mathscr{E}\{P_tP_t'\} = vP_t, \qquad \mathscr{E}\{U_{m,t}U_{m,t}'\} = vU_{m,t},$$
$$\mathscr{E}\{P_tP_{ti+1}'\} = cP_{t,t+1}, \qquad \mathscr{E}\{U_{m,t}U_{m,t+x}'\} = cU_{m,t,t+x},$$

where $Y_{m,t,n}$ = the raw scores on measure m at time t for individual n

$P_{t,n}$ = the common factor scores at time t for individual n

$U_{m,t,n}$ = the unique factor scores measure m at time t

$j_{m,t}$ = the factor loadings for measure m at time t

vP_t = the variance of the common factor at time t

$cP_{t,t+1}$ = the covariance of common factors over time

$vU_{m,t}$ = the unique variance of measure m at time t

$cU_{m,t,t+x}$ = the unique autocovariance of measure m over time delay x

Also sX = the standard deviation of component X

rXY = the correlation of component X and Y

(b) Nonlinear structural model of covariances

$$\mathscr{E}\{vY_{m,t}\} = j_{m,t}\, vP_t\, j_{m,t}' + vU_{m,t},$$
$$\mathscr{E}\{cY_{m,t,t+x}\} = j_{m,t}\, cP_{t,t+x}\, j_{m,t+1}' + cU_{m,t,t+x}, \quad \text{and}$$
$$\mathscr{E}\{cY_{m,m+k,t,t+x}\} = j_{m,t}\, cP_{t,t+x}\, j_{m+k,t+1}'.$$

(c) Parameter estimates

Model parameter	#0 Two-factor baseline	#1 Invariant loading	#2 Equal unique	#3 Equal size	#4 Invariant scores	#5 Random scores
Factor loadings at first occasion (age 6)						
$P_1 \rightarrow Y_{1,1} := j_{1,1}$	1.000	1.000	1.000	1.000	1.000	1.000
$P_1 \rightarrow Y_{2,1} := j_{2,1}$	1.537*	1.049**	.964**	1.024**	1.025**	1.034**
$P_1 \rightarrow Y_{3,1} := j_{3,1}$	1.037*	1.028**	1.014**	1.022**	1.031**	1.031**
$P_1 \rightarrow Y_{4,1} := j_{4,1}$	1.119*	.974**	.926**	.954**	.991**	.933**
Factor loadings at second occasion (age 11)						
$P_2 \rightarrow Y_{1,2} := j_{1,2}$	1.000	1.000	1.000	1.000	1.000	1.000
$P_2 \rightarrow Y_{2,2} := j_{2,2}$	.901*	1.049**	.964**	1.024**	1.025**	1.034**
$P_2 \rightarrow Y_{3,2} := j_{3,2}$	1.021*	1.028**	1.014**	1.022**	1.031**	1.031**
$P_2 \rightarrow Y_{4,2} := j_{4,2}$	.915*	.974**	.926**	.954**	.991**	.922**
Factor score parameters						
$P_1^* \leftrightarrow P_2^* := r_{12}$	.755*	.751*	.801*	.608*	1.00	.000
$P_1^* \rightarrow P_1 := sP_1$	4.47*	4.96*	4.88*	8.18**	4.34*	4.98*
$P_2^* \rightarrow P_2 := sP_2$	10.76*	10.18*	10.78*	8.18**	9.97*	10.33*
Unique factor variances at first occasion (age 6)						
$U_{1,1} \leftrightarrow U_{1,1} := vU_{1,1}$	17.3*	15.8*	30.0**	33.0**	22.3*	14.1*
$U_{2,1} \leftrightarrow U_{2,1} := vU_{2,1}$	47.2*	53.9*	66.0**	60.8**	63.7*	55.0*
$U_{3,1} \leftrightarrow U_{3,1} := vU_{3,1}$	35.5*	34.1*	61.2**	63.3**	42.2*	32.1*
$U_{4,1} \leftrightarrow U_{4,1} := vU_{4,1}$	14.4*	15.6*	20.1**	19.3**	18.6*	18.5*
Unique factor variances at second occasion (age 11)						
$U_{1,2} \leftrightarrow U_{1,2} := vU_{1,2}$	47.0*	49.5	30.0**	33.0**	53.1*	41.8*
$U_{2,2} \leftrightarrow U_{2,2} := vU_{2,2}$	73.5*	71.7	66.0**	60.8**	79.5*	69.3*
$U_{3,2} \leftrightarrow U_{3,2} := vU_{3,2}$	88.8*	91.3	61.2**	63.3**	91.7*	92.7*
$U_{4,2} \leftrightarrow U_{4,2} := vU_{4,2}$	23.9*	23.0	20.1**	19.3**	28.0*	21.7*

(continued)

TABLE 5.7. Continued.

Unique factor covariances between occasions (ages 6 and 11)						
$U_{1,1} \leftrightarrow U_{1,2} := cU_{1,t}$	1.0*	.3*	.1*	−.6*	−4.5*	1.7*
$U_{2,1} \leftrightarrow U_{2,2} := cU_{2,t}$	−.2*	−1.6*	−1.4*	1.0*	−9.7*	.5*
$U_{3,1} \leftrightarrow U_{3,2} := cU_{3,t}$	−1.5*	−2.3*	−1.2*	−3.9*	−6.1*	−1.6*
$U_{4,1} \leftrightarrow U_{4,2} := cU_{4,t}$	1.7*	2.3*	1.6*	1.9*	−2.9*	3.7*
(d) Goodness-of-fit indices						
Model Index	#0 Two-factor baseline	#1 Invariant loading	#2 Equal unique	#3 Equal size	#4 Invariant scores	#5 Random scores
Likelihood ratio	25	40	127	255	106	147
Degrees-of-freedom	15	18	22	23	19	19
Raw Z-test	1.65	2.86	8.0	13.	7.3	9.1
Measure of centrality	97%	95%	77%	56%	81%	73%
Likelihood index change	0%	−60%	−408%	−800%	−324%	−488%
Interpretation	Two factor good	Loadings invariant	Unique unequal	Sizes unequal	Scores change	Scores not random

Horn et al., 1983; Nesselroade and Bartsch, 1977; McArdle, 1988; Olsson and Bergmann, 1977).

This kind of multivariate measurement is useful in many ways. Here we assume that a single factor characterizes the data at each time point. This model has both common and unique factor loadings within each time point, and this leads to several useful features. First, because exactly the same manifest variables Y_t have been measured at each time, we can separate the common covariance over time from the specific covariance over time. This is a useful separation that nicely matches the typical separtion of common and specific variance within each time point. Second, we can estimate the factor loading j_t at each time point, and this provides factor scores $P_{t,n}$, which, in theory, have no measurement error. In each of the previous models we have tried to add some estimates of reliability and unreliability. These attempts were limited by the fact that only one variable was measured at each time. Adding multiple variables allows the possibility of *reliable factor scores,* and this can have an impact on any further change scores analysis.

Another strength of the longitudinal factor model is how it deals with the principle of *factor invariance over time.* The measurement of the same variables by itself does not ensure that the factor loadings j_t are identical. In this model the assumption of identical loadings is a testable hypothesis. The factor loadings reflect a functional patterning of the response Y_t to the stimulus P, and this may change over time. In the prior models change over time was defined by change in level or changes in scatter, but this model adds the possibility of changes in the factors themselves. This is often the most basic question about development and changes. In the repeated-measure case a test of factor loadings, $j_1 = j_2$, allows us to examine the question. "have we measured the same construct at each time point?"

Additional tests of hypotheses about the *invariance of factor scores* can also be examined. Factor scores P_t can be related to one another by an autoregressive effect or as a difference component. As the variance of a factor approaches zero, the factor is said to decay, but if the variance increases, the factor is said to grow. Furthermore, as the variance of a factor approaches its covariance with another factor (and hence, $r_{12} = 1.0$), the factor may be said to collapse. Most criticallly, factor loading invariance is not the same as factor score invariance, and changes can appear at either level. The factor pattern is extrinsic between persons, whereas the factor scores are intrinsic within each person. These issues are often viewed as questions about *qualitative and quantitative* changes, and these questions can be examined simultaneously within this SEM.

3.8. Longitudinal Factor Examples

This kind of model cannot be fitted to the data of Table 5.1. The main problem is that we have listed the statistics for only onv variable—WISC-verbal. In order to illustrate this problem now, we take the scores for four separate measures of the WISC-Verbal: Information, Comprehension, Similarities, and Vocabulary. To simplify this analysis further, we also take these four variables measured at just two occasions, 6 and 11. This forms an 8×8 covariance matrix to be fitted by the structural equation model of Figure 5.7. More details on these data and analysis can be found in McArdle and Nesselroade (1988).

The first model we fit is baseline model #0. This is a two-factor model where we assume one factor at each time point. The loading for the first variable is fixed at $j_1 = 1.0$ to provide a scale for the factor scores, and the rest of the factor loadings are free to vary as ratios of this scaling. The factor at time 1 has loadings $j_1 = \{1.000, 1.537, 1.037, 1.119\}$, and the factor at time 2 has loadings $j_2 = \{1.000, .901, 1.021, .915\}$. These are not identical, so these two factors are not interpretable as the identical construct. The variance of each factor is given by its standard deviation $sP_1 = 4.47$ (and $vP_1 = 19.98$) and $sP_2 = 10.76$ ($vP_2 = 115.8$), and the correlation among the factor scores $r_{12} = .755$. The unique variances vU_t are not assumed to be equal, and each repeated variable is allowed to have a nonzero unique covariance cU. Unlike the prior baseline models, this model fits these data very well with LRC = 25 on DF = 15 and $Z = 1.65$. These LRC and DF are not comparable to the previous models because we are fitting a different covariance matrix.

Model #1 is the model of invariant loadings. Here we obtain factor loadings $j_t = \{1.000, 1.049, 1.028, .974\}$, and these apply to these data at ages 6 and 11. Notice that the factor variances and covariances do not change very much with these additional restrictions. For example, the correlation among the two factor scores is now $r_{12} = .751$ (instead of .755). Nevertheless, this is now correctly interpreted as an *auto*correlation because the factor is now explicitly scored in the same way at each occasion. This model fits the data with LRC = 40 in DF = 18 with $Z = 2.86$. The difference between this model #1 and the two-factor baseline #0 gives dLRC = 15 on dDF = 3 because we have estimated three fewer parameters. Invariance is not much different than the baseline, but it is a 60% loss of fit.

It is useful to restate what this testing of factor invariance implies. If we accept the invariant factor loadings, we are formally stating we have measured the same construct in the same way at each occasion. In this case the construct might be, say, verbal ability, and this is reflected in WISC-Verbal scores in the same way at each time. If so, we can move on and examine other parameters of change. If we reject the hypothesis of equal loadings, we are stating that there has been a more fundamental or qualitative change in the meaning or measurement of the construct. By looking back at the results for model #0, we can see the second variable, WISC-Comprehension, has a higher loading at age 6 than do the other WISC variables, but it also has an almost equal loading at age 11. Thus, we might say that we largely measure verbal comprehension at age 6, but we measure a more equal range of verbal ability at age 11. The "configural" model of change can be interesting and has reasonable empirical properties (see Horn, et al., 1983). However, if this metric characteristic of the factor has changed over time, we cannot easily examine changes in factor means, variances, and correlations, and variation of loadings contradicts our earlier scoring assumptions. These are interesting but often difficult choices to make (for more details see Horn et al., 1983, McArdle, 1988).

The remaining models of Table 5.7 assume the factor loadings j_t are identical over both occasions and examine some characteristics of the factor scores. In model #2 we add the restriction of equal unique variances uV_t across both occasions. This model obtains LRC = 127 on DF = 22, so dLRC = 87 on dDF = 4. Therefore this hypothesis is probably unreasonable. Model #3 adds a restriction that the common factors have equal variance of $sP = 8.18$. The fit of this model is LRC = 255 on DF = 23, so this is a very bad idea. It is clear that this factor of verbal ability increases in variability between ages 6 and 11. Since this variability is the key expression of the size of the factor, we think we have measured more verbal ability at age 11 than at age 6.

The characteristic patterning of the factor scores are important to any model of this kind. Model #4 assumes the invariant loading and adds a fixed value of the autocorrelation $r_{12} = 1.0$. This is a pure stability model where the factor scores are invariant over both occasions. This model fits with LRC = 106 on DF = 19 and, compared to model #1, the test of factor score stability gives dLRC = 66 on dDF = 1, so this extreme hypotheses is unacceptable. Model #5 takes this hypotheses in the other direction and forces $r_{12} = 0$, so the factors are uncorrelated over time. If this model fits, the factor scores on the identical constructs are entirely unrelated. This model fits with LRC = 147 on DF – 19, so dLRC = 107 on dDF = 1, and this is one of our worst ideas.

The last few models illustrate the power of this organization for the analysis of change. These models demonstrate how the questions of invariance of factor scores are mathematically and statistically independent from questions about invariance of factor loadings. We can examine factor score invariance without factor pattern invariance, and vice versa. The final model with an invariant factor pattern with zero factor correlations is probably our most extreme case. Nevertheless, this can be a useful organization for a model of state factors (see Hertzog and Nesselroade, 1987) or it can be a feature of longitudinal data where individuals change at differing rates (see Wohlwill, 1988). These distinct structural hypotheses are often confused in analyses of factor stability. The results of this model suggests the idea of factor loading invariance is not such a bad starting point for the organization of changes in uniqueness, increasing factor variance, and we have some positive correlation of the factor over time. More details on these analyses appear in McArdle and Nesselroade (1988).

4. Discussion

The previous structural equation models represent different ideas about change. Yet, as we have illustrated, each model can be fitted to the same basic set of data. The direct comparison of these models of change is potentially informative. However these are simply illustrations of more general structural equation principles. We now deal with a few of these more general issues.

4.1. Change Models and Extensions

We have used latent variable structural equation models to provide a formal patterning for several different models of change or development. In the autoregression model we examined change as a process that comes from a stable background and a series of independent influences. In the difference components model we examined change as a series of independent increments that build up (or decay) and persist over time. In the growth curve models we viewed change as a developmental process that unfolded or spurted ahead over time. In the longitudinal factor model we examined change as a feature of the measurement process and as a feature of the scores.

The structural equation approach allowed us to formalize alternative change concepts within each of these models and to use the same mathematical and statistical techniques to estimate and evaluate these models. This is a nontrivial benefit because we can now combine many aspects of these change models together. For example, we can estimate a latent growth curve model with autoregression among the error components. Or we can estimate a longitudinal factor analysis model with latent growth among the factor scores. In more complex combinations we can have latent growth in many variables and examine the factor pattern of these change variables. The addition of other path analysis variables or multiple groups adds several new dimensions to this problem. Some examples of these kinds of combinations have been presented by Joreskog (1979), Olsson and Bergmann (1977), McArdle and Epstein (1987), and McArdle (1988).

Longitudinal data are obviously important, and their benefits are widely heralded. More advanced autoregressive and cross-lagged models are complex but potentially informative, especially when age- and time-related changes in the parameters are investigated (Horn and McArdle, 1980). The latent growth curve SEM capitalizes on other benefits of longitudinal data. These models concentrate on the study of individual differences in changes over age and use time and age as dimensions upon which these changes are evaluated. These longitudinal extensions of the difference components models come from a much older tradition in psychological research on learning (see Nesselroade, 1972; Rogosa and Willett, 1985), but they are on the rise again in SEM (Meredith and Tisak, 1984, this volume; McArdle, 1986, 1988). Perhaps more than any other SEM, these longitudinal models follow the proposals of Wohlwill (1973, 1988).

There are many other extensions of these basic change ideas. In life-span development we usually encounter mixed forms of longitudinal and cross-sectional data. This means *convergence* or *missing data* models are needed, and multiple group longitudinal structural equations models are likely to be a continued necessity (e.g., Meredith and Tisak, 1984; McArdle, Anderson, and Aber, 1987). Other interindividual comparisons, such as the randomized groups design, are usually indicated as well because statistical assumptions are always critical to the determination of univariate or multivariate effects on developmental trajectories (see McArdle, 1988b). Another important grouping of individuals is based on biometric heritage (e.g., monozygotic and dizygotic twins, adopted and biological families, etc.). This leads to the separation of genetic and nongenetic sources of behavioral growth variables and broad developmental trajectories (e.g., McArdle, 1986). It is also useful to recognize the developmental significance of individualized single -subject designs; i.e., the use of a P-technique approach (see Cattell, 1966; Nesselroade and McArdle, 1986; Wohlwill, 1988). All these ideas have been represented as other patterns of change within SEMs.

4.2. Current Disadvantages of SEM

There are many disadvantages to current SEM techniques that we should not overlook. Some of these limitations emerge from basic assumptions about the raw data required and other limitations deal with mathematical and statistical models we routinely use.

Most SEMs require *large and representative samplings of subjects.* Large sample sizes are required to have the power to distingush between large and complex models, and sample size can have an effect of model selection (see Matsueda and Bielby, 1986; Marsh, Balla and McDonald, 1988). No doubt, simple models with small numbers of parameters will require few subjects, and more complex models with large number of parameters will require larger number of individuals or replicates. But, and perhaps more important, even when working with advanced latent variable models the individuals need to be independent and representatively sampled. Some aspects of nonrandom subject selection can be handled within SEM, but these are limited to homogeneous and linear effects. Nonrandom subject selections where the groups have different models cannot be handled by regular SEM techniques (but see Woodbury and Manton, 1983). If we measure many subjects from a minor submodel of a theory, but fewer subjects are measured for the major submodel of a theory, then the latent factors and parameters will reflect the minor subpopulation. If subjects of a special type are not measured then there is no data to estimate the latent parameters. These old messages of regression analysis apply with great force in SEM.

Still another SEM limitation is the need for *large and representative sampling of variables.* But, and again more importantly, even when we are working with advanced latent variable models, the critical variables need to be measured in a suitably representative fashion. If many variables appear for the minor factors of a theory, but fewer variables appear for the major factors of a theory, then the latent factors and parameters will largely reflect the minor factors. If some variables are not measured, then there is no data to estimate the latent parameters. Testable hypotheses about a latent construct require at least four good indicators. In longitudinal data the number of time points as well as the spacing between the time points needs to be considered as a selection of marker variables from a domain of interest. These old messages of factor analysis apply with great force in SEM (Horn and McArdle, 1980).

Statistical modeling issues have been carefully documented in the SEM literature, but we have only alluded to these problems in the previous results. Standard SEMs seem to require *symmetry of the dependent variables*. Recent work on more general data distributions described by Browne (1984) is helpful, but it is not intended to be a corrective for all problems. For example, dependent variables with skewed distributions can lead to serious problems for later SEM. This poses a real problem for categorical response data. In binary item data, for example, the variance is strictly determined by the difficulty level. Needless to say, these special cases require special consideration before simply applying SEM techniques (see McDonald, 1983, 1987; von Eye, 1988).

A related limitation is the *linearity of the response function*. If the variable scoring system has not been organized to linearize the relationship among the variables, the SEM will not produce proper linear analysis. There are many ways to examine nonlinearity in data by transformations or reparameterizations. In some models we can rescore the original data by using a power transformation Y^P of the response variable (e.g., the Box-Cox transformation; Draper and Smith, 1981) and use these new scores in further SEM analysis. But most current SEM programs do not provide any facility for estimating the value of power P from the manifest data Y.

Another related limitation is based on the *intrinsic linearity of the model*. There are many functional relationships of interest between ages and scores that are impossible to translate to a linear form with additive residuals. In the aging literature the exponential and Gompertz forms are often used to define growth and decline, and these cannot be linearized with equal error (Draper and Smith, 1981). Functions that can be reparameterized into linear models can be fitted with SEM techniques, but those that cannot require complex scalar devices (available in COSAN).

Some of these limitations are created by the reflexive adherence on standard SEM programs, such as LISREL and COSAN. These programs require data to be summarized as first and second moments, and these statistics are seriously limited by categorical measurement, heterogeneity, and subject self-selection effects. One general way to account for these problems is to expand the computational model to be fitted to the data matrix. In this approach each subject adds only some parameters to the overall model, and the likelihood is built up over N independent groups. This technique was originally proposed by Wilks (1932) in his theory of fragmentary data, and implemented by Little and Rubin (1987). Incidentally, a similar approach has been used in behavioral genetic work on complex pedigrees by Jardine and Martin (1984).

The standard SEM programs are overly complicated and relatively expensive to run, and we need to make these better. Certainly any move toward a more complex N-group analysis adds substantial computational burdens. The recent upsurge of expectation-maximization (EM) algorithms can be important here. These EM algorithms estimate surrogates for latent variable scores at each iterative step and complex problems are optimized in partitions (see Little and Rubin, 1987; Tisak and Meredith, 1987). These and other computational approaches point out problems with standard SEM analyses and represent the future of better SEM.

4.3. Patterns of Change within SEM

The main benefit of SEM comes because it allows us to formalize different kinds of change concepts. SEM makes it easier to represent and test change concepts, and this makes it easy to communicate, think about, and compare different change ideas. This makes it possible, for example, to use SEM to contrast standard "change in parameters" models with "parameters of change" models. A diverse range of change concepts can be formalized and studied with this mathematical perspective.

The main limitation of SEM is illustrated when the mathematical statements, by pictures or by numbers, are taken too seriously. The formalisms conveyed by simple pictures have a tendency to oversimplify and mislead; form takes the place of function. Many knowledgeable researchers have jumped at the simple-minded assumption of causality from an arrow projecting ahead in time in a SEM diagram. The pictures create illusions that are hard to deny, so it is easy to overstate causality or to overstate the benefits of maximized likelihoods. Often we are just proud that the computer program is finally finished. Yet, it is surprising that the initial path model by Sewell Wright (1921; Figure 5.1) was a far more complex idea than most causal models in press today. The search for *invariant parameters of change,* a goal suggested by Wright (1921), is still difficult (see Goldberger, 1973; Nesselroade, 1983).

Perhaps the most important lesson we can convey is that there are different models of change and *each model tests explicit hypotheses given implicit assumptions.* These structural models help make some of these assumptions more explicit, and this may help us understand what we can and should say about our data This also means there is no single best way to organize change data. As our recent history shows, models that match the zeitgeist of one era can turn into an anachronism during another. The choice of an appropriate change model is based on a relatively complex series of substantive and methodological issues. The statistical features of maximum likelihood estimation and goodness-of-fit indices are rigorous and appealing, but these are far less important than the initial organization of an appropriate mathematical model of change. The best part of the recent work in structural equation models has provided one useful way to clarify, communicate, and pattern these changes.

A1. Technical Appendix

There are many technical details of the methods we have used that deserve further detail. Most of these details can be found in the treatments of McDonald (1985), Loehlin (1987), or Joreskog and Sorbom (1979, 1985). In these last few tables (5A.1) to (5A.3) we define a few useful features of the SEM approach described earlier.

A1.1. General Structural Equations

In this presentation we have used a particular algebraic model for the definition of a structural prediction system. This aproach is described in detail by McArdle and McDonald (1984), and some of these ideas are summarized in Table 5A.1. In Table 5A.1(a) we list the linear models specification for any set of linear model equations. The key elements of this model are vectors v, g, and u, and matrices A, S, and F. In Table 5A.1(b), we list the nonlinear expectations required for the calculation of the structural expectations. The key elements in this equation are the geometric expansion of A into effect matrix E, the connections $C = ESE'$ and the manifest connections $Cg = FCF'$. It is useful to recall that this algebra is related to a generalized set of graphic principles, and these are detailed in McArdle and Horn (1988).

These equations can be directly used for the estimation and testing of the model parameter estimates. In Table 5A.1(c) we define the residuals R_{ij} for a particular set of connections C_{ij} (from the example used in the

Table 5A.1. General Equations for a Linear Structural Equation Model

(a) Linear model specification

$$v = Av + u, \quad g = Fv, \quad \text{and} \quad \mathscr{E}\{uu'\} = S,$$

where $v = 1 \times t$ vector of all internal model variables

$u = 1 \times t$ vector of all external model variables

$g = 1 \times j$ vector of observed variables

$A = t \times t$ matrix of asymmetric coefficients (arrows)

$S = t \times t$ symmetric matrix of symmetric coefficients (slings)

$F = j \times t$ permutation matrix of fixed elements (filters)

(b) Nonlinear matrix equation

Given

$$v = Av + u, \quad \mathscr{E}\{uu'\} = S, \quad \text{and} \quad g = Fv,$$

then

$$v - Av = u, \quad v(I - A) = u, \quad v = (I - A)^{-1}u,$$

$$v = Eu, \quad \text{where } E = (I - A)^{-1}, \quad \text{and}$$

$$E = (I - A)^{-1} = A^0 + A^1 + A^2 + A^3 + \cdots + A^r,$$

where E = the $t \times t$ effect matrix of order r (length r).

If $\mathscr{E}\{vv'\} = C$, then

$$C = \mathscr{E}\{[\mathscr{E}u][Eu]'\} = \mathscr{E}\{Euu'E\} = E\mathscr{E}\{uu'\}E',$$

and $C = ESE'$, where C = the $t \times t$ connection matrix of all model variables.
If $\mathscr{E}\{gg'\} = Cg$, then

$$g = FEu,$$

and

$$Cg = \mathscr{E}\{[FEu][FEu]'\} = FE\mathscr{E}\{uu'\}E'F' = FESE'F',$$

or

$$Cg = FCF',$$

where

Cg = the $j \times j$ expected matrix of connections among observed variables.

(c) Nonlinear structural equations for the simple example

$$C_{21} = R_{21} + az = R_{21} + a(C_{11})$$

$$C_{32} = R_{32} + bh1h' + baza' = R_{32} + b(C_{22})$$

$$C_{43} = R_{43} + cj1j' + cbh1h'b' + cbaza'b' = R_{43} + c(C_{33})$$

$$C_{31} = R_{31} + bcz = R_{31} + ab(C_{11})$$

$$C_{42} = R_{42} + cbh1h' + cbaza' = R_{42} + bc(C_{22})$$

$$C_{41} = R_{41} + cbaz = R_{11} + abc(C_{11})$$

$$C_{11} = R_{11} + z = R_{11} + z$$

$$C_{22} = R_{22} + h1h' + aza' = R_{22} + (aza' + hh')$$

TABLE 5A.1. Continued.

$$C_{33} = R_{33} + j1j' + bhh1h'b' + baza'b' \quad = R_{33} + b(aza' + hh')b' + jj'$$

$$C_{44} = R_{44} + k1k' + cj1j'c' + cbh1h'b'c' = R_{44} + c(b(aza' + jj')b')c' + kk' + cbaza'b'c'$$

where C_{ij} = observed covariances

$P = \{a, b, c, h, j, k, z\}$ = parameters to be estimated

R_{ij} = residual covariances to be minimized (see appendix)

(d) Likelihood ratio criteria for cross-products expectations

$$F(ml) = \text{LRC} = N\{\text{LRD} + \text{LRT} + \text{LRM}\}, \quad \text{so}$$

$$F(ml) \approx \sum_{i=1}^{M} \sum_{j=1}^{i} [O\{i, j\} - E\{i, j\}]^2 / E\{i, j\}^2,$$

where

$O\{i, j\} = \Sigma\,(Y(i)Y(j))/N$ = observed values of score products

$X\{i, j\} = \mathscr{E}\{Y(i)Y(j)'\}$ = expected values of score products

$\text{LRD} = \ln|O\{C\}| - \ln|\mathscr{E}\{C\}|$ = likelihood ratio based on log determinants

$\text{LRT} = \text{tr}\{O(C)\mathscr{E}\{C\}^{-1}\} - t$ = likelihood ratio based on matrix trace

$\text{LRM} = (O\{M\} - \mathscr{E}\{M\})\mathscr{E}\{C\}^{-1}(O\{M\} - \mathscr{E}\{M\})'$ = likelihood based on means (T^2)

(e) Numerical approximation (see McDonald, 1978)

$$P(i+1) = P(i) + D(i) \quad \text{and} \quad D(i) = G(i)H(i)^{-1}a,$$

where at the ith iteration,

$P(i) = \text{vec}[A(i), S(i)]$ = the vector of free parameters

$D(i) = \text{vec}[\partial A(i), \partial S(i)]$ = the vector of parameter changes

$G(i) = \text{vec}[\partial A(i)/\partial F(i), \partial S(i)/\partial F(i)]$

= the vector of first-order derivatives of the parameters with respect to the function (i.e., the gradient vector),

$H(i) = \text{vec}[\partial A(i)\, \partial A(i)/\partial^2 F(i), \partial A(i)\, \partial S(i)/\partial^2 F(i), \partial S(i)\, \partial S(i)/\partial^2 F(i)]$

= the vector of second-order derivatives of the parameters with respect to the function (i.e., the hessian matrix),

a = a scalar used to avoid divergence.

(f) Additional goodness-of-fit indices

$$Z = \{[\text{LRT}/DF]^{1/2} - [1 - [2/9]/DF\}/[[2/9]/DF]^{1/3}$$

= null model normal Z-score (see Horn and McArdle, 1980).

$\text{LIP} = \text{LRC}(m)/\text{LRC}(0)$ = likelihood change as a percentage.

$Mc = \exp(-.5(\text{LRT}/N - DF/N)) \times 100$ = McDonald's noncentrality index.

After McArdle and McDonald 1984.

Chapter). In Table 5A.1(d) we define the likelihood ratio criterion for a general matrix of cross-products (for further elaboration see McArdle, 1986, 1988). In Table 5A.1(e) we define the matrix calculus required to take an arbitrary set of parameters and change these to conform with an explicit set of data, a model, and a minimization function. In Table 5A.1(f) we list formulas for some goodness-of-fit indices used to our statistical analyses.

A1.2. LISREL and COSAN Computer Program Specification

The statistical estimation of the parameters of any linear model can be done with any available algorithm for numerical optimization—the canned packages of LISREL (Joreskog and Sorbom, Note 1, 1979), COSAN (McDonald, 1978, Fraser and McDonald, 1988), the optimiza-

TABLE 5A.2. LISREL Input for Age-Based Autoregressive Model (see Table 5.4)

```
TI  Full Equal Autoregression—McArdle and Epstein (1987, Child Development)
DA NG = 1   NI = 6   NO = 204   MA = CM
LA
     'WISC_6','WISC_7','WISC_9','WISC_11','MotherEd','Constant'
ME
  18.034  25.819  35.255  46.593  10.811  1.000
SD
   6.374   7.319   7.796  10.386   2.700   .000
KM SY
  1.000
   .809  1.000
   .806   .850  1.000
   .765   .831   .867  1.000
   .520   .531   .448   .458  1.000
   .000   .000   .000   .000   .000  1.000
SE
  1  2  3  4  /
MO NY = 4   NE = 11   BE = FI, FU   PS = FI, SY   LY = FI   TE = ZE
LE
  '1.WISC_6','2.WISC_7','3.WISC_8','4.WISC_9','5.WISC10','6.WISC11',
  '7.ERR_7','8.ERR_8','9.ERR_9','10.ERR_10','11.ERR_11'
MA LY
  1  0  0  0  0  0  0  0  0  0  0
  0  1  0  0  0  0  0  0  0  0  0
  0  0  0  1  0  0  0  0  0  0  0
  0  0  0  0  0  1  0  0  0  0  0
```

TABLE 5A.2. Continued.

```
MA BE
 0  0  0  0  0  0  0  0  0  0  0
 1  0  0  0  0  0  1  0  0  0  0
 0  1  0  0  0  0  0  1  0  0  0
 0  0  1  0  0  0  0  0  1  0  0
 0  0  0  1  0  0  0  0  0  1  0
 0  0  0  0  1  0  0  0  0  0  1
 0  0  0  0  0  0  0  0  0  0  0
 0  0  0  0  0  0  0  0  0  0  0
 0  0  0  0  0  0  0  0  0  0  0
 0  0  0  0  0  0  0  0  0  0  0
 0  0  0  0  0  0  0  0  0  0  0
MA PS
 1
 0  0
 0  0  0
 0  0  0  0
 0  0  0  0  0
 0  0  0  0  0  0
 0  0  0  0  0  0  1
 0  0  0  0  0  0  0  1
 0  0  0  0  0  0  0  0  1
 0  0  0  0  0  0  0  0  0  1
 0  0  0  0  0  0  0  0  0  0  1
FR BE(2, 1) BE(2, 2) BE(4, 3) BE(5, 4) BE(6, 5)
FR BE(2, 7) BE(3, 8) BE(4, 9) BE(5, 10) BE(6, 11)
FR PS(1, 1)
EQ BE(2, 1) BE(3, 2) BE(4, 3) BE(5, 4) BE(6, 5)
EQ BE(2, 7) BE(3, 8) BE(4, 9) BE(5, 10) BE(6, 11)
OU RS TV PT NS SS ND = 3
```

tion library, or NAG (Eaves et al., 1978). The LISREL algorithms have become widely available and have proven successful in numerous applications (e.g., Horn and McArdle, 1980; McArdle, 1986b), so we provide an example of LISREL programming here in Table 5A.2.

Any structural equation model can be directly specified in LISREL:

- All given variables are assigned a unit value in "LY" (e.g., "START 1.0 LY(1,1)"), and this totally fixed FILTER (F) matrix can be totally ignored on output.
- All ARROWs are parameters from column to row in "BE" (e.g., "FREE BE(2,1)"), and this ASYMMETRIC (A) matrix may be constrained in any way (e.g., "MODEL BE = IN").

- All SPANs are parameters among column and row in "PS" (e.g., "FIX PS(2,2)"), and this SYMMETRIC (S) matrix may be constrained in any way (i.e., "MODEL PS = PS").
- Degrees of freedom corrections for any fixed unit constant moments is simply done by subtraction to the output degrees of freedom.

The advantage to programming LISREL with RAM notation is that it is clear to start a model, easy to change a model, and it produces accurate results. This direct RAM specification of LISREL V can result in large and sparse RAM matrices F, A, and S, For example, any model that includes four manifest variables, one unit constant, and one common factor, requires a statement to LISREL with the number of LISREL manifest variables ("Y") is "NY = 5" and the number of LISREL latent variables ("ETA") is "NE = 6". The average cross-products matrix can be practically evaluated with the LISREL V algorithm by either (1) externally calculating the cross-products and using the "MA = CM" convention or (2) augmenting a manifest unit constant to the data, reading in the means and covariances, and using the "MA = MM" convention.

The direct RAM is not a computationally efficient way to run LISREL. Further data-based and LISREL-based specifications (e.g., the "No-X" model) could be introduced for further computational speed, but these are generally less didactic, and they are certainly hard to draw. Models moving away from RAM and more toward LISREL can also be quite costly in terms of human programming time, so they are not suggested.

The models described here may also be programmed with the COSAN-II program (Fraser and McDonald, 1988). To program any RAM model in COSAN, the steps may be simply outlined as follows:

- Specify a COSAN model with order $m = 2$.
- Specify the RAM-F matrix as COSAN F(1).
- Rewrite the RAM-A matrix as (A-I) and set into COSAN F(2) and use the inverse option B(2).
- Specify the RAM-S matrix as the COSAN P matrix.

This simple RAM specification is accurate, but it does not take advantage of the most useful features of COSAN logic. For example, one of the advantages in using COSAN-II is that the expectations for the third- or fourth-order factor model do not require matrix inversion at all. This is theoretically different from LISREL, which requires inversion for any third- or higher-order model. Once again, only two or three COSAN programs need to be written to accommodate the problems described here.

A1.3. RAMIT Computer Program Specification

The RAM matrices are usually large and sparse, so it can be convenient to program with a sparse matrix notation. We have developed a computer algorithm specifically designed to deal with these issues. The RAMIT program is discussed by McArdle (1986b, 1988) and by McArdle and Horn (1988). Basically, RAMIT is a new set of language options built around the path graphic concepts. We do not provide full details here, but we do offer one sample of RAMIT input in Table 5A.3.

Table 5A.3 gives an example of the RAMIT input for the same model as defined to LISREL in Table 5A.2. In RAMIT we use a parser called MC, which reads in a series of characters and number up to the

TABLE 5A.3. RAMIT Input for Age-Based Autoregressive Model (see Table [5.4])

```
/* This RAMIT input uses a List structure and uses minimal LISREL based input.
  A variety of other options are possible. */
        set_title = "Full Equal Autoregression from McArdle & Epstein (1987)";
        set_density = 204;
        set_print = most;

input_data
        standard_deviation, 1, 1, 6.374;
        standard_deviation, 2, 1, 7.319;
        standard_deviation, 3, 1, 7.796;
        standard_deviation, 4, 1, 10.386;
        standard_deviation, 5, 1, 2.700;
        correlation, 2, 1, .809;
        correlation, 3, 1, .806;
        correlation, 3, 2, .850;
        correlation, 4, 1, .765;
        correlation, 4, 2, .831;
        correlation, 4, 3, .867;
        correlation, 5, 1, .520;
        correlation, 5, 2, .531;
        correlation, 5, 3, .448;
        correlation, 5, 4, .458;
        ;
input_model
        filter, 1, 1, 1, 0, 0, 0, "wisc-6";
```

(continued)

TABLE 5A.3. Continued.

```
        filter, 2, 2, 1, 0, 0, 0, "wisc-7";
        filter, 3, 4, 1, 0, 0, 0, "wisc-9";
        filter, 4, 6, 1, 0, 0, 0, "wisc-11";
        arrow, 2, 1, .5, .1, 1, 0, "6 to 7";
        arrow, 3, 2, .5, .1, 1, 0, "7 to 8";
        arrow, 4, 3, .5, .1, 1, 0, "8 to 9";
        arrow, 5, 4, .5, .1, 1, 0, "9 to 10";
        arrow, 6, 5, .5, .1, 1, 0, "10 to 11";
        arrow, 2, 7, .8, .1, 2, 0, "error at 7";
        arrow, 3, 8, .8, .1, 2, 0, "error at 8";
        arrow, 4, 9, .8, .1, 2, 0, "error at 9";
        arrow, 5, 10, .8, .1, 2, 0, "error at 10";
        arrow, 6, 11, .8, .1, 2, 0, "error at 11";
        sling, 1, 1, 1, .1, 0, 0, "initial scaling";
        sling, 7, 7, 1, 0, 0, 0, "fixed scaling";
        sling, 8, 8, 1, 0, 0, 0, "fixed scaling";
        sling, 9, 9, 1, 0, 0, 0, "fixed scaling";
        sling, 10, 10, 1, 0, 0, 0, "fixed scaling";
        sling, 11, 11, 1, 0, 0, 0, "fixed scaling";

        set_method = maximum-likelihood;
        set_iterations = 100;
        estimate_model;

output_model;

output_data;

set_save;
        file = "trueauto.out";

set_end;
```

semicolon delimiter. This means we can define INPUT, OUTPUT, ESTIMATE, and SET as *functions* that can be called anytime, and we can then read the appropriate values into the appropriate memory locations. Several other RAMIT options that are similar to options used in the earlier LISREL program are not detailed here.

We include Table 5A.3 because it defines a specific way to describe the input and output of a structural equation model. In this input stream we use a *parameter list* L, where each row is defined as

$$L_{(p)} = \{M_{(p)}, R_{(p)}, C_{(p)}, V_{(p)}, S_{(p)}, E_{(p)}, X_{(p)}\},$$

where p = the pth parameter in the model, $M_{(p)}$ = the matrix or type of the parameter (i.e., 1 = arrow, 2 = span), $R_{(p)}$ = the row variable of the parameter (1 to t), $C_{(p)}$ = the column variable of the parameter (1 to t), $V_{(p)}$ = the value of the parameter (or symbolic labels), $S_{(p)}$ = the standard error of the parameter (or symbolic labels), $E_{(p)}$ = any equality constraint on the parameter, and $X_{(p)}$ = some additionally useful information about the parameter.

The length of the list depends on the total number of parameters P in a model or in the data; the elements within the list may all be numeric or symbolic, and the rows within the list are arbitrary. When needed for algebraic computations, this list is used to reconstruct the required RAM matrices F, A and S.

Each model parameter can be read directly off the list (e.g., the first parameter is "an arrow to variable r_1 from variable c, which is termed b_1 and is fixed.) The filter parameters are used to designate which variables in the list are manifest. The list above is organized into arrows, spans, and filter elements, but the row organization is completely arbitrary. For calculation purposes it will be useful to have an entirely numerical list also. Again each model parameter can be read directly off the list (e.g., the first parameter is "a one-way relation to variable 1 from variable 9, which has a value of 1.00 and an error of 0.00"). The standard error is a key feature of the model: For the pth parameter, $SE_{(p)} > 0$ imply p is "free," $SE_{(p)} = 0$ imply p is "fixed," but a nonzero value in the sixth column implies p is "equal" to some other parameter in the list. The setting of equality constraints by a single list pointer allows an arbitrary row ordering of the list.

The list notation above represents a simple and convenient programming device for specifying any kind of path model. The RAMIT program performs an interface function automatically: (1) The RAM list structure is used on input. (2) The RAM list is converted to a COSAN list. (3) Calculations are carried out using COSAN (without inversion if possible). (4) the COSAN results are stored back in the original RAM list order. This interface allows maximal COSAN programming efficiency without any user translation, and this logic forms the basis of the RAMIT programs. These program conventions simply reflect current knowledge in this area. Hopefully, RAMIT conventions will allow change at a rapid rate.

References

Belsley, D. A., Kuh, E., and Welsch, R. E. (1980). *Regression diagnostics: identifying influential data and sources of collinearity*. New York: Wiley.

Blalock, H. M. (Ed.) (1985). *Causal models in panel and experimental designs.* New York: Aldine.

Browne, M. W. (1984). "Asymptotically distribution-free methods for the analysis of covariance structures," *British Journal of Mathematical and Statistical Psychology.* **37,** 62–83.

Cattell, R. B. (1966). *Handbook of multivariate experimental psychology.* New York: Rand McNally.

Cleveland, W. S. (1985). *The elements of graphing data.* Monterey: Wadsworth.

Cook, T. D., and Campbell, D. T. (1979). *Quasi-experimentation: design and analysis issues for field settings.* Chicago, Ill.: Rand McNally.

Cronbach, L. J., and Furby, L. (1970). "How should we measure "change"—or should we?" *Psychological Bulletin.* **74,** 68–80.

Cudeck, R., and Browne, M. W. (1983). "Cross-validation of covariance structures," *Multivariate Behavioral Research.* **18,** 147–167.

Draper, N. R., and Smith, H. (1981). *Applied regression analysis,* 2nd ed. New York: Wiley.

Duncan, O. D. (1969). "Some linear models for two wave, two variable panel analysis," *Psychological Bulletin.* **72,** 177–182.

Eaves, L. J., Last, K. A., Young, P. A., and Martin, N. G. (1978). "Model fitting approaches to the analysis of human behavior," *Heredity,* **41,** 249–320.

Fraser, C., and McDonald, R. P. (1988). COSAN: "Covariance structure analysis," *Multivariate Behavioral Research.* **23,** 263–265.

Goldberger, A. S. (1973). "Structural equation models: an overview." In A. S. Goldberger and O. D. Duncan (Eds.), *Structural equation models in the social sciences.* New York: Seminar Press.

Gollob, H. F., and Reichardt, C. S. (1987). "Taking account of time lags in causal models," *Child Development.* **58,** 80–92.

Guttmann, L. (1954). "A new approach to factor analysis: the radex." In P. F. Lazarsfeld (Ed.), *Mathematical thinking in the social sciences* (pp. 258–348). Glencoe, Free Press.

Hertzog, C., and Nesselroade, J. R. (1987). "Beyond autoregressive models: Some implications of the trait-state distinction for structural equation modeling of developmental change," *Child Development.* **58,** 93–109.

Horn, J. L. (1972). "State, trait and change dimensions of intelligence," *The British Journal of Educational Psychology.* **42**(2), 159–185.

Horn, J. L., and McArdle, J. J. (1980). "Perspectives on mathematical/statistical model building (MASMOB) in research on aging." L. W. Poon (Ed.), *Aging in the 1980's: selected contemporary issues in the psychology of aging* (pp. 503–541). Washington, D.C.: American Psychological Association.

Horn, J. L., McArdle, J. J., and Mason, R. (1983). "When is invariance not invariant? a practical scientist's look at the ethereal concept of factor invariance," *The Southern Psychologist.* **1**(4), 179–188.

Hultsch, D. F., Nesselroade, J. R., and Plemons, J. K. (1976). "Learning ability relations in adulthood," *Human Development.* **19,** 234–247.

Jardine, R. and Martin, N. G. (1984). "No evidence for sex-linked or sex-limited gene expression influencing spatial orientation," *Behavior Genetics.* **14**(4), 345–354.

Jones, M. B. (1962). "Practice as a process of simplification," *Psychological Review.* **69**(4), 274–294.

Joreskog, K. G. (1970). "Estimation and testing of simplex models, "*British Journal of Mathematical and Statistical Psychology.* **23,** 121–146.

Joreskog, K. G. (1979). "Statistical estimation of structural models in longitudinal-developmental investigations." In J. R. Nesselroade and P. B. Baltes (Eds.), *Longitudinal research in the study of behavior and development.* New York: Academic Press.

Joreskog, K. G. and Sorbom, D. (1979). *Advances in factor analysis and structural equation models.* Cambridge, Mass. Abt Books.

Joreskog, K. G., and Sorbom, D. (1985). *LISREL-VI program manual.* Chicago: International Educational Services.

Kenny, D. A. (1979). *Correlation and causality.* New York: Wiley.

Little, R. T. A., and Rubin, D. B. (1987). *Statistical analysis with missing data.* New York: Wiley.

Loehlin, J. C. (1987). *Latent variable models: An introduction to factor, path, and structural analysis.* Hillsdale, N. J.: Lawrence Erlbaum.

Marsh, H. W., Balla, J. R., and McDonald, R. P. (1988). "Goodness-of-fit indices in confirmatory factor analysis: the effect of sample size," *Psychological Bulletin.* **103**(3), 391–410.

Matsueda, R. L., and Bielby, W. T. (1986). "Statistical power in covariance structure models." In N. B. Tuma (Ed.), *Sociological methodology 1986* (pp. 120–158). San Francisco: Jossey-Bass.

McArdle, J. J. (1980). "Causal modeling applied to psychonomic systems simulation, *Behavior Research Methods and Instrumentation.* **12,** 193–209.

McArdle, J. J. (1986a). "Latent growth within behavior genetic models," *Behavior Genetics.* **16**(1), 163–200.

McArdle, J. J. (1986b). "RAMIT: A computer program for structural equation modeling." Paper presented at the Annual meeting of the *Society of Multivariate Experimental Psychology.* Atlanta, Georgia. October 1986.

McArdle, J. J. (1988). "Dynamic but structural equation modeling with repeated measures data." In J. R. Nesselroade and R. B. Cattell (Eds.), *Handbook of multivariate experimental psychology,* Vol. II. New York: Plenum Press.

McArdle, J. J. (1988b). "Structural models of developmental theory in psychology. In *Annals of theoretical psychology,* Vol. VI (In press).

McArdle, J. J., Anderson, E., and Aber, M. S. (1987). "Convergence hypotheses modeled and tested with linear structural equations." *Proceedings of the 1987*

Public Health Conference on Records and Statistics (pp. 347–352). National Center for Health Statistics, Hyattsville, Md.

McArdle, J. J., and Epstein, D. (1987). "Latent growth curves within developmental structural equation models," *Child Development.* **58,** 110–133.

McArdle, J. J., and Horn, J. L. (1988). "An effective graphic model for linear structural models, *Multivariate Behavioral Research.* (In press).

McArdle, J. J., and McDonald, R. P. (1984). "Some algebraic properties of the Reticular Action Model for moment structures," *British Journal of Mathematical and Statistical Psychology.* **37,** 234–251.

McArdle, J. J., and Nesselroade, J. R. (1988). "Structuring data to study development and change." In S. H. Cohen and H. W. Reese (Eds.), *Life span developmental psychology: methodological innovations.* Hillsdale, N. J.: Lawrence Erlbaum (in press).

McDonald, R. P. (1978). "A simple comprehensive model for the analysis of covariance structures," *British Journal of Mathematical and Statistical Psychology.* **31,** 59–72.

McDonald, R. P. (1980). "A simple comprehensive model for the analysis of covariance structures: some remarks on applications," *British Journal of Mathematical and Statistical Psychology.* **33,** 161–183.

McDonald, R. P. (1983). "Linear and nonlinear models in item response theory," *Applied Psychological Measurement.* **6,** 379–396.

McDonald, R. P. (1985). *Factor analysis and related methods.* Hillsdale, N. J.: Lawrence Erlbaum. (In press).

McDonald, R. P. (1984). "Describing the elephant: structure and function in multivariate data," *Psychometrika.* **51**(4), 513–534.

Meredith, W., and Tisak, J. (1984). " "Tuckerizing" curves." Paper presented at the Annual Meetings of the Psychometric Society, Santa Barbara, Calif.

Nesselroade, J. R. (1972). "Note on the 'longitudinal factor analysis' model," *Psychometrika.* **37**(2), 187–191.

Nesselroade, J. R. (1977). "Issues in studying developmental change in adults from a multivariate perspective." In J. E. Birren and K. W. Schaie (Eds.), *Handbook on the psychology of aging* (pp. 59–69). New York: Van Nostrand Reinhold.

Nesselroade, J. R. (1983). "Temporal selection and factor invariance in the study of development and change." In P. B. Baltes (Ed.) *Life-span development and behavior,* Vol. 5, (pp. 59–87). NY: Academic Press.

Nesselroade, J. R., and Bartsch, T. W. (1977). "Multivariate perspectives on the construct validity of the trait-state distinction." In R. B. Cattell and R. M. Dreger (Eds.), *Handbook of modern personality theory.* New York: Wiley.

Nesselroade, J. R., and McArdle, J. J. (1986). "Multivariable causal modeling in alcohol use research," *Social Biology.* **32**(3–4), 272–296.

Olsson, U., and Bergman, L. (1977). "A longitudinal factor model for studying change in abilities," *Multivariate Behavioral Research.* **12,** 221–241.

Osborne, R. T., and Suddick, D. E. (1972). "A longitudinal investigation of the intellectual differentiation hypothesis, *Journal of Genetic Psychology.* **110,** 83–89.

Rao, C. R. (1958). "Some statistical methods for the comparison of growth curves," *Biometrics.* **14,** 1–17.

Rogosa, D. and Willett, J. B. (1985a). Understanding correlates of change by modeling individual differences in growth," *Psychometrika.* **50**(2), 203–228.

Short, R., Horn, J. L., and McArdle, J. J. (1983). "Mathematical and statistical issues in the analysis of developmental data." In R. N. Emde, and R. J. Harmon (Eds.), *Continuities and discontinuities in development* (pp. 371–401). New York: Wiley.

Sobel, M. E., and Bohrnstedt, G. W. (1985). "Use of null models in evaluating the fit of covariance structure models." In N. B. Tuma (Ed.), *Sociological methodology, 1985.* San Francisco, Calif.: Jossey-Bass.

Spearman, C. (1904). "General intelligence objectively determined and measured," *American Journal of Psychology.* **15,** 201–293.

Tisak, J., and Meredith, W. (1989). "Exploratory longitudinal factor analysis in multiple populations," *Psychometrika,* **54,** 261–281.

Tucker, L. R. (1958). "Determination of parameters of a functional relation by factor analysis," *Psychometrika.* **23**(1), 19–23.

Tucker, L. R. (1966). "Learning theory and multivariate experiment: illustration by determination of parameters of generalized learning curves." In R. B. Cattell (Ed.), *The handbook of multivariate experimental psychology* (pp. 476–501). Chicago: Rand-McNally.

von Eye, A. (1988). *Configural frequency analysis,* Department of Human Development, Pennsylvania State University.

Wilks, S. S. (1932). "Moments and distributions of estimates of population parameters from fragmentary samples," *Annals of Mathematical Statistics.* **3,** 163–195.

Wohlwill, J. F. (1970). "The age variable in psychological research," *Psychological Review.* **77**(1), 49–64.

Wohlwill, J. F. (1973). *The study of behavioral development.* New York: Academic Press.

Wohlwill, J. F. (1988). "The partial isomorphism between developmental theory and methods," *Annals of theoretical psychology,* Vol. VI (In press).

Wold, H. (1949). "A large sample test for moving averages," *Journal of the Royal Statistical Society* (*Series B*). **2,** 297–305.

Woodbury, M. A., and Manton, K. G. (1983). "A mathematical model of physiological dynamics of aging and correlated mortality selection. I. Theoretical development and critiques," *Journal of Gerontology.* **38**(4), 398–405.

Wright, S. (1921). "Correlation and causation," *Journal of Agricultural Research*. **20,** 557–585.

Wright, S. (1934). "The method of path coefficients," *Annals of Mathematical Statistics*. **5,** 161–215.

Wright, S. (1984). "On 'Path analysis in genetic epidemiology: A critique'," *American Journal of Human Genetics*. **35,** 757–768.

Chapter 6 Applications of Scaling to Developmental Research*

PHILLIP WOOD

Psychology Department
University of Missouri-Columbia
Columbia, Missouri

Abstract

Many theories of development and change involve strong assumptions of individual differences in growth and change not found in many traditional statistical methodologies. Nonmetric multidimensional scaling techniques are proposed as one family of techniques currently underutilized by developmental psychologists. Nonmetric scaling solutions generate maps of the underlying stimulus patterns, which make fewer assumptions of the data than is often the case in statistical model construction. Additionally, some models construct graphic representations of systematic subject differences. The basic graphic analogy underlying metric scaling solutions is outlined, along with the three fundamental assumptions of distance in scaling. Stress measures for determining the relative fit of candidate nonmetric solutions to the data are presented. Two applications from developmental psychology are presented that illustrate the application of nonmetric scaling solutions. The first, a weighted Euclidean analysis of data on children's conservation of volume, incorporates a specific model of interindividual differences as well as a model of intraindividual

* This chapter was supported by National Institute of Aging Grant T21-AG00-110-02. The author wishes to acknowledge the helpful comments of Alexander von Eye and Mark Davison on earlier drafts of this paper. The author also wishes to acknowledge the support of the Max-Planck Institute for Human Development and Education, Berlin, Federal Republic of Germany.

variation. The second example concerns a nonmetric analysis of developmental change in ratings of children's behavior, and determines optimal levels of environmental characteristics for individuals. This example has particular relevance to developmental researchers analyzing behavior from a system theoretic point of view. In this external nonmetric analysis, ideal optimal models are easily computable, given common regression analysis software. Common problems generally encountered in conducting nonmetric scaling analysis, local minima, degenerate solutions, and lack of convergence are defined, and courses of action to prevent their occurrence are proposed.

Scaling methods may be broadly defined as statistical techniques developed to study relationships between stimuli presented to subjects at the inter- and intraindividual level. In this respect factor analysis, cluster analysis, and discriminant analysis qualify as scaling techniques under this definition. Although recent developments in comparison of scaling and its more parametric counterparts such as factor analysis reveal many close parallels, which will be discussed, the concentration in this chapter will be on models that either make less restrictive assumptions of the data or that incorporate models that more closely conform to certain assumptions of a developmental perspective of growth and change.

In recent years, scaling techniques have been applied to a wide variety of areas, including counseling psychology (Davison, Richards and Rounds, 1986), personality (Rosenberg and Jones, 1972), and speech development (Shepard, 1972). In spite of this range of applications, metric and nonmetric multidimensional scaling has remained a relatively unused technique in developmental assessment. This is odd, given the fact that several metric models allow for simultaneous estimation of intra- and interindividual change processes. Complaints about the theoretical restrictions and lack of applicability of many standard linear models that assume the normal distribution are common in the developmental literature (Nesselroade and Ford, 1987; Labouvie and Nesselroade, in press; Nesselroade and Cattell, 1989). Scaling alternatives are attractive because several of them do not require the multivariate normality of tests frequently employed. This chapter presents the techniques of scaling as an often overlooked set of techniques that can be applied to data analyzable by other techniques mentioned here, as well as by types of data not amenable to the common methods of analysis due to limitations imposed by the specific design of an experiment, or theoretical assumptions about the nature of the phenomenon under study.

In one sense, attempting to write a single chapter on scaling applications is as difficult as writing a single chapter on the possible developmental applications of the linear model. To complicate matters, scaling models have been articulated by several researchers who independently developed their models. For this reason, early scaling models often used very different terminology to describe similar processes—a confusing development that hides hierarchical relationships between models and confuses the beginning student. For this reason, this chapter will use the terminology of Young (1987) and Davison (1983), because these model designations highlight the explicit measurement assumptions of each model by reference to its geometric representation.

The scaling techniques discussed here produce a geometric or spatial representation of relationships between stimuli or individuals. The end goal of the analyses is to produce a geometric representation of the underlying individual and/or stimulus dimensions that fit the data reasonably well and that can be plotted. Central to all scaling techniques discussed here is the notion of spatial proximity between pairs of objects as a measure of their relatedness. Some measures of proximity found in developmental research include correlation coefficients, joint probabilities, conditional probabilities, and transitional probabilities. Data gathering techniques currently underutilized can also be analyzed by scaling models. These include direct ratings of perceived (dis)similarity between stimuli that come from presenting subjects with sets of pairwise comparisons of a defined stimulus set or from asking parents or children the percent of times that a given strategy of behavior results in a criterion behavior.

In many respects the development of a proximity measure is limited only by the ingenuity of the researcher. A necessary correlate of this statement is that the researcher must be able to defend his or her choice of a proximity measure as reasonable and conforming to the assumptions of the scaling technique employed. These assumptions relate to the precise manner in which proximity (or, to be more precise, lack of proximity) conforms to geometric analogy underlying the model. The explicit choice of the type of distance (Euclidean or non-Euclidean) assumed in the analysis will be understood within the generalized distance function. The way in which models that make increasingly less restrictive assumptions regarding how observed data relate to the underlying latent dimensions will then be presented.

The best way to appreciate scaling alternatives in developmental psychology is to go through examples of analysis. To this end, two analyses of developmental data will be presented. The first example will discuss an individual differences model of perception in a developmental study. The second will involve assessment of a developmental analysis of behavior from a systems perspective.

1. Dissimilarity as Distance: Assumptions of the Latent Structure

Scaling measurement models have most often begun from Richardson's (1938) statement of the relation of psychological dissimilarity (not similarity) and the geometric concept of distance. In order for a set of stimuli to be modeled on this concept of distance, any pair of objects (a, b) must have a latent distance function (d) that meets the following four axioms. We will discuss each one in turn and give a few comments about the degree to which some developmental data may fail to meet the last two of these assumptions.

1. $d[a, a] = 0$. The distance between any object and itself must be zero.

2. $d[a, b] \geq 0$. The distance between the two objects is greater than or equal to zero.

3. $d[a, b] = d(b, a]$. This commutativity axiom states that the distance from a to b is the same as the distance from b to a. Although this would seem to be a reasonable requirement for many data sets, there is evidence that some psychological phenomena do not meet this requirement. Experiments that attempt to assess the degree to which letters presented to subjects are confused with each other is one example of such a process. In such situations, for example, the probability of confusing Q with O is not the same as the probability of confusing O with Q. A dissimilarity measure based on the proportion of times that a given letter is confused with another may fail to meet this assumption. As another example, the probability that a child is able to crawl, given that he or she can speak, is not equal to the probability that the child can speak, given that she or he can crawl. There are two basic ways out of these difficulties. One is to respecify the variables under analysis. For example, the researcher may wish to concentrate on analysis of joint probabilities instead of conditional probabilities and may be able to eliminate the violation of the assumption in this manner. There also exist versions of scaling that do not require that the input matrix be symmetric. These methods are beyond the scope of this chapter, and the reader is referred to Young's (1987; Young and Lewyckyj, 1979) treatment of the analysis.

4. $d[a, b] + d[b, c] \geq d(a, c]$. This is known as the *triangle inequality* and is taken from the geometric relationship stating that when three points lie on a line the sum of line segment AB and line segment BC is equal to line segment AC and is greater than line segment AC for all other cases. This is the most problematic of the four scaling analogies. Nothing about what we know of psychological processes suggests that dissimilarity measures should satisfy this. Given the fact that many scaling models do not make the researcher assume multivariate normality, and the fact that many types of ipsatively scored measures can be analyzed with scaling in order to arrive at nonipsative underlying dimensions, this has seemed like a worthwhile trade-off. As Davison (1983, p. 3) remarked, somewhat tongue in cheek, "In the social and behavioral sciences, however, three axioms out of four is not bad, so the [distance] analogy caught on."

2. Distance Metric

The Minkowski distance function, often used in scaling models, is the general underlying model of latent distance functions. In scaling in more than one dimension, stimulus estimates are points in a coordinate system. Distances d_{ij} are expressed as

$$d_{ij} = \left(\sum |x_{ik} - x_{jk}|^{p} \right)^{1/p},$$

where $x_{..}$ is the stimulus position, i and j are two stimuli, and k is the latent dimension. The choice of p must be positive. Different values of p correspond to different types of distance. For $p = 2$, the distance is the common Euclidean distance; that is, distances between stimuli can be thought of as straight lines between each other. For $p = 1$, the distance is a city-block model (Attneave, 1950), which got its name because in two dimensions the distance between points i and j corresponds to the distance one would travel along city streets if those streets ran through the stimulus points and parallel to the coordinate axis. As p approaches infinity the distance function conforms to a dominance metric in which the distance between stimuli i and j is determined by the difference between coordinates along only one dimension, that dimension for which $|x_{ik} - x_{jk}|$ is the largest. Although no studies employing dominance metrics in developmental assessments are known to this author, the dominance metric may be useful for developmental researchers who believe that subjects may be attending to only the dimension along which stimuli are the most salient.

In the presence of no compelling theoretical reasons for particular choice of p, which p to use was one of the early difficult choices facing scaling applications. Arnold (1971) and Weiner-Ehrlich (1978) used the overall fit of the model in their choice of p. Sherman (1972) suggests that this is a misleading strategy, however. By and large, researchers in the area seem to have adopted a pragmatic solution to their choice of p. Computation problems of local minima, lack of convergence, and degeneracy of solutions occur more frequently in solutions with metrics other than 2 (the Euclidean model). A large amount of software is available for doing scaling, which assumes the Euclidean model (Systat, SAS, MLSCALE, KYST, to name just a few), and the Euclidean model is much easier to interpret to a larger research community.

As we will see in the two experimental examples given here, Euclidean distance is a useful, readily interpretable metric to adopt, which parsimoniously explains relationships in the data. Given research that assumes that children may attend to one and only one dimension of a particular stimulus to the disregard of remaining stimulus characteristics, developmentalists may also wish to consider the dominance metric as representing the underlying dimensions that children use in understanding stimulus materials.

3. Assumptions about the Data

The discussion of the four distance axioms of the latent dimensions and the choices of metric for the latent dimensions says nothing of the relationship of the *observed* dissimilarities δ_{ij} to the true latent distances between stimuli. The required assumptions of the relationship of the observed stimuli to latent dimensions on which those stimuli differ has become less and less restrictive over the history of scaling. Torgerson's (1952) original work on scaling assumed that the obtained dissimilarity data was at least proportional to the underlying distances. Subsequently, Shepard's (1962) model assumed (except for measurement and sampling error) that the relationship of the data to the underlying function (assuming the Euclidean model) is

$$\delta_{ij} = f[d_{ij}] - f\left[\sum (x_{jk} - x_{ik})\right]^{1/2},$$

where f is a strictly increasing monotonic function of d_{ij}. This means that

$$d_{ij} > d_{i'j'} \Rightarrow \delta_{ij} > \delta_{i'j'}.$$

Kruskal (1964) extended this work to all distance models that conform to the general Minkowski space. Nonmetric extensions based on Guttman's (1968) work were proposed by Lingoes (1973), which are in many ways analogous to Kruskal's.

It is possible to describe two broad classes of scaling. *Metric multidimensional scaling* describes those models that assume Torgerson's (1952) direct relationship between observed measurements and the underlying latent dimensions. *Nonmetric multidimensional scaling* describes models that assume Kruskal's (1964) or Shepard's (1962) extensions of the model.

4. Models of Individual Differences: Weighted Euclidean Models

The first class of nonmetric models that we will consider employs the Euclidean model but acknowledges that there exist systematic individual differences in the model. In developmental psychology in the Piagetian tradition, the manner in which subjects interpret multiple attributes of stimulus materials has been an important focus of research. One example of such research is the acquisition of the notion of the conservation of volume. In such studies, subjects are presented two identical glasses of water. The contents of one glass is then poured into a third glass. This glass may have a large diameter or a small diameter. Subjects are then asked which glass has more water in it. Frequently, children who have not yet achieved the level of formal operations will fail to attend to both the height and diameter of the glasses used and will say that the glass with the higher level of water has more water in it. Other children may attent to only the diameter of the glass and say that the glass with the greater diameter has more water in it in spite of the fact that the levels of water in each glass may be quite different. Children who possess the concept of conservation of volume are able to make correct judgments about the relative amounts of water in the glass by simultaneously incorporating information on both the height of water in the glass and the diameter of the glass.

In a slight variation on this paradigm, the study here presented children with glasses of water of varying sizes and with water filled to varying heights. Children were asked to mark, on a six-point scale how different they judged the two glasses of water to be. The stimuli in the study were composed of glasses of diameters 1, 2, and 3 inches and filled to levels of 1, 2, and 3 inches. Four subjects were presented all possible pairs of the combinations of the glasses. Table 6.1 shows the dissimilarities for each stimulus combination for all subjects. Each stimulus is designated G_{ij}, where i is the height of the stimulus and j is its width. Note that all numbers are positive: The children were not asked to estimate which glass contained more water, but were merely asked to rate how different the contents of the two glasses were from each other.

The different patterns of response are quite striking across the children. Note that for child B, the only dimension that seems to be attended to is the dimension of width, since stimuli differing in height (G_{13}, G_{23}, and G_{33}, for example) all received a value of 0, indicating no difference between the glasses. Although the dissimilarity ratings of child A seem related to the differences in volume across the stimuli, note that sometimes the child fails to discriminate between differences beyond a given magnitude. For example, the comparison of G_{12} and G_{33} results in a dissimilarity rating of 5, and the comparison of G_{22} and G_{33} also results in a dissimilarity rating of 5. Such a difference may be the result of a ceiling effect on the six-point scale used, or it may reflect the lack of sensitivity in the subject's "guesstimate" of how different the stimuli were. Although the estimate is not sensitive to the differences in magnitude, the dissimilarity rating can be thought of as monotonically related to the underlying dimension of volume.

One of the first tasks in analysis of such data using nonmetric scaling is the choice of the number of dimensions to represent the relationships between stimuli. As a general rule, this choice is the same as in factor analysis: If a conceptual rationale can be made for the number of dimensions in the data, this is preferable to computational methods to determine the number of dimensions in the data. In factor analysis, an eigenvalues-by-factors plot is frequently used to determine the dimensionality of the solution. In nonmetric scaling, an eigenvalue estimate is not available. The adequacy of a solution depends in part on the interpretability of the obtained solution and the degree to which the nonmetric solution accounts for the data. To determine how many dimensions most parsimoniously account for the data, nonmetric analyses are run in several dimensions. The stress values that result are then plotted by the dimensions in the various solutions to determine the appropriate dimensionality.

Table 6.1. Dissimilarity Ratings of Glasses of Water of Varying Heights and Diameters for Four Children

Stimulus	G_{13}	G_{23}	G_{33}	G_{12}	G_{22}	G_{32}	G_{11}	G_{21}	G_{31}	Stimulus	G_{13}	G_{23}	G_{33}	G_{12}	G_{22}	G_{32}	G_{11}	G_{21}	G_{31}
Child *A*										Child *C*									
G_{13}										G_{13}									
G_{23}	3									G_{23}	3								
G_{33}	5	3								G_{33}	5	3							
G_{12}	1	4	5							G_{12}	0	3	5						
G_{22}	1	2	5	2						G_{22}	3	0	3	3					
G_{32}	3	0	3	4	2					G_{32}	5	3	0	5	3				
G_{11}	2	5	5	1	3	5				G_{11}	0	3	5	0	3	5			
G_{21}	1	4	5	0	2	4	1			G_{21}	3	0	3	3	0	3	3		
G_{31}	0	3	5	1	1	3	2	1		G_{31}	5	3	0	5	3	0	5	3	
Child *B*										Child *D*									
G_{13}										G_{13}									
G_{23}	0									G_{23}	3								
G_{33}	0	0								G_{33}	1	3							
G_{12}	3	3	3							G_{12}	1	4	5						
G_{22}	3	3	3	0						G_{22}	1	2	5	2					
G_{32}	3	3	3	0	0					G_{32}	3	0	3	3	1				
G_{11}	5	5	5	3	3	3				G_{11}	2	3	5	1	3	5			
G_{21}	5	5	5	3	3	3	0			G_{21}	1	3	5	1	1	1	1		
G_{31}	5	5	5	3	3	3	0	0		G_{31}	1	2	5	1	1	1	2	1	

4.1. Stress Measures

There are basically five stress measures of fit in nonmetric scaling. The two sets of stress values developed by Kruskal were developed to minimize distances between coordinate points in the estimated solution and the estimated disparities between stimuli assuming the model. Kruskal's first measure of stress, *stress formula 1* (S_1) is defined as follows:

$$S_1 = \left| \frac{\sum_{(i,j)} (\delta_{ij} - \delta'_{ij})^2}{\sum_{(i,j)} \delta'^2_{ij}} \right|^{1/2},$$

where d'_{ij} is the estimated distance between stimulus i and stimulus j under the estimated solution, and δ'_{ij} is the estimated disparities between the stimuli assuming the model. The assumed metric in the model we are going to employ for these data is Euclidean. The d'_{ij} are therefore represented as

$$d'_{ij} = \left(\sum_{(k)} |x'_{ik} - x'_{jk}|^2 \right)^{1/2},$$

where k is the number of dimensions in the solution.

The δ'_{ij} are parameters chosen so as to be close to the distance estimates as possible and still monotonically related to the original dissimilarity data.

Stress formula 2 (S_2) is like S_1 except for a different denominator;

$$S_2 = \left| \frac{\sum_{(i,j)} (\delta_{ij} - \delta'_{ij})^2}{\sum_{(i,j)} (\delta'_{ij} - d'_{..})^2} \right|^{1/2}$$

where

$$d'_{..} = \frac{(\sum_{(i,j)} d'_{ij})}{IJ}$$

In other words, $d'_{..}$ is the average estimated distance.

Alscal (Young and Lewyckj, 1979), a very versatile nonmetric program, minimizes a slightly different stress formula called *S-stress formulas 1 and 2.* These are identical to stress formulas 1 and 2, except that they minimize the squared distances and disparities instead of only the distances and dissimilarities; that is,

$$SS_2 = \left| \frac{\sum_{(i,j)} (\delta^2_{ij} - \delta'^2_{ij})^2}{\sum_{(i,j)} \delta'^2_{ij}} \right|^{1/2}$$

and S-stress formula 2 (SS_2) is

$$SS_2 = \left| \frac{\sum_{(i,j)} (\delta_{ij}^2 - \delta_{ij}'^2)^2}{\sum_{(i,j)} (\delta_{ij}'^2 - d_{..}'^2)^2} \right|^{1/2}.$$

A final variant in computed stress is used in Multiscale (Ramsey, 1986). Ramsey cites convergence desiderata for the log-likelihood function of the stress formula as defined by the sum of squares error (SSE) given as

$$\mathrm{SSE} = \sum_r \sum_{(i,j)} (f_r \log d_{ijr} - \log d_{ijr}^*)^2$$

Nonmetric scaling programs generally minimize one of these stress formulas. Some programs, such as KYST, give the user a choice in which stress formula to use in estimating scale values. On the whole, when the data matrix is a symmetric matrix of dissimilarities, some of the computational problems discussed here are not as frequent under S_1 (Kruskal and Carroll, 1969; Roskam, 1969; Takane, Young, and DeLeeuw, 1977). If the data are preference data, however, S_2 leads to fewer problems.

A final measure of stress in nonmetric scaling programs is called the *coefficient of alienation*. This coefficient is defined as $(1 - u^2)^{.5}$, where u is the coefficient of monotonicity (a measure of the degree of ordinal association between the original data and the distance estimates) defined as

$$u = \frac{\sum_{(i,j)} (\delta_{ij}' d_{ij}')}{(\sum_{(i,j)} \delta_{ij}'^2 \sum_{(i,j)} d_{ij}'^2)^{.5}}$$

Lingoes's (1973) smallest space analysis (SSA) programs give coordinate estimates that minimize the coefficient of alienation. For additional points of interpretation of the coefficient, see Lingoes (1973).

4.2. *Stress-by-Dimensions Plot*

In the conservation data employed here, SS_1 was used. Nonmetric scaling solutions for dimensions 1 through 5 were computed and are graphed in Figure 6.1. As can be seen, there is a marked elbow at two dimensions, meaning that the two-dimensional solution most parsimoniously accounts for the relations in the data set. In data sets where the dimensionality of the data are not readily apparent from the stress-by-dimensions plot, the researcher can compare the interpretability of the solutions generated,

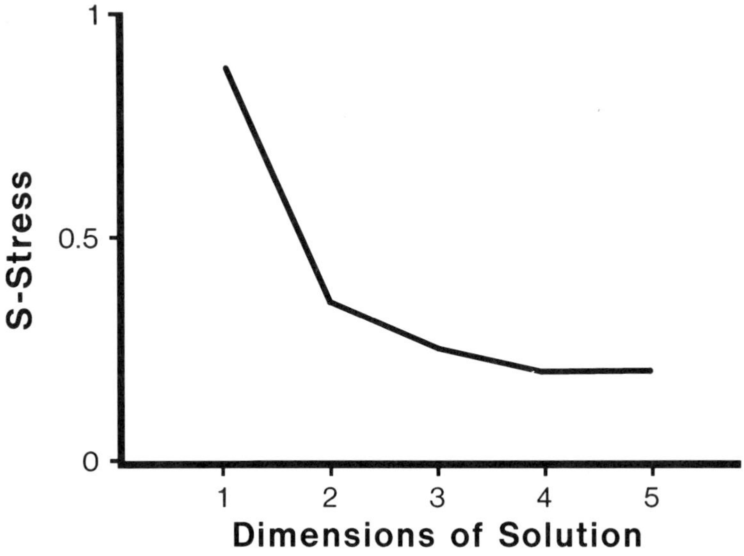

FIGURE 6.1. Stimulus by dimensions plot.

settling on a configuration that most closely corresponds to the conceptual framework of the study. In situations where the stimuli can be rated on the basis of an external criterion, the researcher may wish to consider an external analysis as described here for ratings of children's behavior.

5. Weighted Euclidean Analysis Results for Sample Data

Based on the fit-by-dimensions plot, two dimensions were to be retained in the example data. The weighted Euclidean nonmetric scaling results for these data are two matrices. The first matrix, the stimulus matrix, shows the underlying characteristics that subjects are using in their ratings of dissimilarity. This 2 (dimensions) $\times$ 9 (stimuli) matrix is graphically represented in Figure 6.2. As can be seen, the obtained solution nicely reproduces the patterns of level of water and the diameter of the glasses. The glasses with width 3 are at the left side of the plot, glasses with dimension 2 are at the middle, and glasses with dimension 1 are at the right. The second dimension can be readily interpreted as a height dimension, with stimuli at the top of the plot representing glasses of height 1, the middle row composed of glasses of height 2, and the bottom row composed of glasses of height 3.

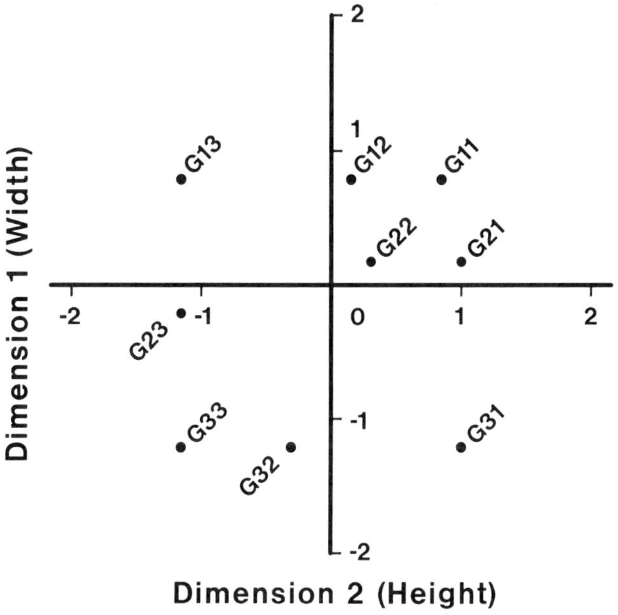

FIGURE 6.2. Stimulus coordinates for sample data.

The second matrix generated by the weighted Euclidean model contain subject weights. For each subject, a coefficient is produced that is a measure of the relative salience of this dimension for a given subject. The subject weights or the four subjects of this study are presented in Table 6.2. In the right-hand columns are shown the S-stress and multiple R^2 for each subject. In this analysis, subject weights were constrained to be

TABLE 6.2. Subject Weights for the Two-Dimensional Weighted Euclidean Model

	Dimension			
Subject	1	2	R^2	S-Stress
1	.47	.45	.43	.26
2	.98	.02	.96	.11
3	.02	.98	.98	.08
4	.43	.40	.35	.28

positive. Although it is an option in many scaling programs to allow negative subject weights, there are often interpretational difficulties associated with them.

Patterns of individual differences across subjects are striking, based on examination of the subject weights. First of all, children A and D appear to have about equal weights on both the dimensions of height and weight. Subjects B and C, though, have very large weights on one but not on the other dimension. Child B appears to be using mainly the dimension of width in judging the dissimilarity of the stimuli. Child C appears to only take into account the height of the stimuli in arriving at decisions of how different the stimuli are. Young (1987) defines a coefficient called *weirdness,* which describes the degree to which a subject has weights on only a few of the dimensions in the overall solution. Subjects who have weights proportional to the average weights across all individuals have a weirdness of 0. Subjects who have only one positive weight have a weirdness of 1. The weirdnesses for children A and D are .01 and .03, respectively. The weirdnesses of subjects B and C are both .97.

Last, although both children A and D have about equal weights on dimensions 1 and 2, the data from child D fit the model less well than from child A. On inspection of the matrix of dissimilarities, we can see that child D's performance differs from child A's on some of the cases where the width of the stimulus was greater (e.g., G_{31} with G_{13} or G_{33} with G_{13}). This could indicate a less consistent pattern of comparison for this subject or a not yet consistent ability to conserve volume.

The analysis presented here given this particular instrument has several advantages over other types of statistical analyses. So long as the dimensions that underlie subjects' judgments are the same, the weighted Euclidean model will furnish a global model that describes the relations between stimuli and the degree to which these dimensions are salient for a given subject. The data neatly summarize the underlying dimensions present, from extraction of the relationships across all individuals, and communicate how a given individual is different from this global pattern. The degree to which the individual fails to conform to this pattern applied to all individuals is evident in the stress values for a given individual. An individual's differential weighting on the assumed dimensions does not, on the whole, swamp the structure of the overall solution.

This analysis makes it possible to identify individuals who develop in a nonnormative manner, indicating that the overall model of how a major group of individuals functions does not apply at all to them. If a child, for example, responded randomly to the stimuli or along entirely different stimulus characteristics, this would result in poorer fit statistics for that child. The researcher so alerted to such problems in summarizing the child's response could then conduct separate analyses on the child or children in order to see if an entirely different pattern of relationships between stimuli attained for the child, or whether no pattern of relationships could be found. It is precisely this type of nonnormative pattern of growth that, in developmental research, is sometimes difficult to identify.

6. Developmental Analysis of Assessments of Behavior: Models of Ideal Points

Although the recovering of underlying dimensions of cognition has a rich tradition within developmental psychology, other developmental researchers have sought to understand the actions of the child within the context of her or his environment. These researchers frequently use the language of systems theory to describe relationships between the individual and the environment (Nesselroade and Ford, 1987; Ford 1987, von Eye and Kreppner, 1989). One of the central focuses of this theorizing is the notion of "optimal level" of a variable to produce a desired outcome. This concept assumes that the individual acts as a self-constructing living system in the environment in order to produce some desired (steady) state of response. This assumption of the relationship of a level of a variable to a response is largely unexplored within the tradition of standard hypothesis testing in psychology. Nesselroade and Ford (1987) have summarized several of the methodological implications of a systems approach for hypothesis testing and the implications of this approach for hypothesis testing. They have also outlined some of the serious difficulties that a researcher adopting a systems approach encounters when attempting to test systems relationships using standard classical hypothesis testing.

It helps to understand the nature of these relationships to consider a concrete example. Suppose, for instance, one is interested in investigating the effect of amount of parent verbalization on whether a child performs a given task. Figure 6.3 shows the relationship between verbalization and probability of acceptable performance for three children. For children A and C, the relationships between verbalization and probability of successful behavior are monotone functions. For child C, the more the parent verbalizes, the greater the probability that the child

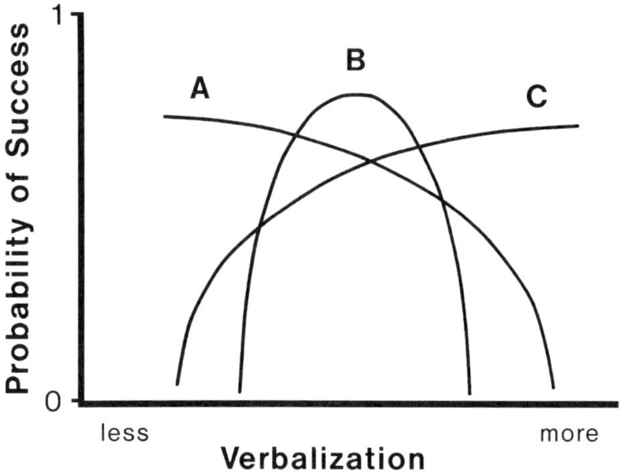

FIGURE 6.3. Probabilities of compliance for three children.

will perform successfully. For child A, the more verbalization, the smaller the probability that the child will perform successfully. If all individuals performed in this fashion, several statistical techniques within the classical and scaling traditions exist for scaling such data [i.e., Tuckerizing of curves (Tucker, 1960), unidimensional scaling (Torgerson 1952; Davison and Wood 1983), factor analysis (Meredith and Tisak, Chapter 4), regression, etc.]. Child B, however, demonstrates a relationship between verbalization and probability quite different from the other two. For this child, there exists an optimal level of verbalization that is best suited to this child. More verbalization is better than less verbalization up to a certain point, at which less verbalization becomes more effective than more. Any parent of an adolescent can report that finding this optimal level of verbalization is not always easy, and that, beyond a certain point, increasing verbalization frequently reduces the probability of success. In addition, the overall pattern of effectiveness of a strategy may be task-specific. The postulate of such a curvilinear relationship between an experimentally manipulated variable and response is one possible operationalization of an optimal level that the individual as a self-constructing living system seeks to maximize.

Notions of optimal level of response, the existence of such large and structural differences between individuals, and the notion individuals as self-constructing systems represent the theoretical underpinnings of several developmental approaches, but are unfortunately not often reflected in the analytic techniques of the discipline. For many situations, the assumption of linear relationships between a predictor and a response variable is thought a simpler and more parsimonious explanation of behavior.

Unfortunately, even in studies that involve repeated measurements of behavior, the types of data that developmental psychologists can gather are often not amenable to the statistical engines that might provide information about the presence or absence of such relationships. Several available models under configural frequency analysis, for example, require gathering data across several individuals, which results in a lumping of interindividual variation across individuals that may obscure the presence of different ideal points for various individuals. Although the configural frequency approach could be extended to assessments of children across many testings, this approach is beyond the resources of a study.

To show how multidimensional scaling might be profitably used to identify if ideal points exist for several individuals, what those ideal points might be, and the relative impact of such variables on the response of children, it is most helpful to go through an example. Consider a fictional study in which mothers are instructed to rate how likely children are to respond to a request to put their toys away. Let's assume that the researcher is interested in assessing the role of relative amount of verbalization as well as the amount of physical contact in the interaction with the child at the time. In our measure, let's assume that verbalization can assume one of four categories; no verbalization to the child, saying a few words to the child, saying a whole sentence to the child, and repeatedly asking the child to put away toys. Suppose that physical contact has the following levels: not touching the child, touch involving the digital or palmar contact by the parent, and embracing the child. The possible behaviors that can result are summarized in Table 6.3. That of no verbalization is, by definition, a *nonevent* or a situation where a request of the child cannot be made and which the mothers will not be asked about. This omission is similar to creating a "structural zero" in the matrix of responses in log-linear modeling (Feinberg, 1980).

Table 6.3 shows generated data for such a study for five such children. Inspection of the data reveals that the probability of the desired behavior in child A is highest for the cell (3, 2), indicating that touching the child with the hands and asking the child in a full sentence is the most effective strategy (probability of success = .60). The least effective strategy is embracing the child but not verbalizing the request ($p = .30$). For child B, however, the highest assigned probability is in cell (2, 2). In addition,

TABLE 6.3. Estimated Proportions of Successful Response for Touch and Verbalization Conditions

Verbalization	Touch: No touch (1)	Hand contact (2)	Embrace (3)
Child *A*			
No verbalization (1)		.4	.3
Phrases/words (2)	.4	.5	.4
Sentence (3)	.5	.6	.5
More than a sentence (4)	.4	.5	.4
Child *B*			
No verbalization (1)		.6	.4
Phrases/words (2)	.5	.7	.5
Sentence (3)	.4	.6	.4
More than a sentence (4)	.3	.5	.3
Child *C*			
No verbalization (1)		.4	.5
Phrases/words (2)	.4	.5	.6
Sentence (3)	.5	.6	.7
More than a sentence (4)	.6	.7	.8
Child *D*			
No verbalization (1)		.2	.1
Phrases/words (2)	.3	.4	.3
Sentence (3)	.5	.6	.5
More than a sentence (4)	.3	.4	.3

the decreases in probability of success seems to decrease more rapidly across levels of touch (decrease between any columns = .2) than for child A. This indicates that differences in the touch dimension are more salient for child A than for child B. For child C, the highest proportion of responses is found in cell (4, 3), indicating that the more touch and the more verbalization the child received in this study, the more likely the child was to perform successfully. Child D's responses are similar to child A's, except that the differences along the verbal dimension seem to be larger than for the touch dimension.

In this application, we will employ what is called an *external preference model* to the data. We will assume that the higher the estimated proportion of success in the child, the more the child prefers the stimulus.

We will further assume the underlying dimensions of the response are known and correspond to the levels as defined in Table 6.3. In order to provide convenient estimates of the ideal points for each child, we will also assume that numerical values of the categories have a continuous and interval interpretation; that is, two verbalizations are somehow twice as much verbalization as one verbalization. These assumptions are merely so that a model can be used that is easily understood. Nonmetric internal analyses can be conducted when the underlying dimensions of interest are now known in the study. The reader may wish to take these data as input to an internal individual differences model similar to the first example presented here and rotate the resulting dimensions to an interpretable solution. See Coombs, McClelland, and Coombs (1973) for an example of such an analysis. The reader seeking a basic introduction to the models is referred to Davison (1983, Chap. 5) and, for a detailed example, to Davison (1983, pp. 152–158). Metric internal analyses are conducted in the same way as the nonmetric examples described above, but assume Torgerson's metric.

Carroll (1972) proposed four models that can account for patterns of preferences in a multidimensional setting. For each of these models, we will discuss the assumptions of the model, how this can be computed using a standard multiple regression statistical package, and the statistical tests appropriate to testing whether the obtained model differs from the null hypothesis of no preference model, as well as comparison of each of the five types with each other.

6.1. Vector Model

The first of these models is called the *vector model.* This model was first proposed by Tucker (1960) and elaborated by Bechtel (1976) and Bechtel, Tucker, and Chang (1971). It assumes that the subject's preference for a given stimulus will increase or decrease linearly with the levels of the stimulus. In Figure 6.3, children A and C have preference curves that satisfy the vector model. Under this model, the observed preferences for subject s on stimulus i, δ_{is}, have the following form (except for measurement error):

$$\delta_{is} = \sum_k b_{ks} + c_s$$

Estimation of the b_{ks} and c_s is easy if the data for a given individual are rearranged as in Table 6.3. In this table, the known dimensions of response are organized by level. Estimation of the b_{ks} and c_s is easy if the data for a given individual are rearranged as in Table 6.3. In this table, the known dimensions of response are organized by level. The columns indicate degree of touch and rows indicate degree of verbalization separately for each child. The probabilities of a successful response from the child is given in the entries of the table. To perform a vector model analysis, the researcher simply performs separate multiple regressions for each child, taking the probability of response as a dependent variable, with the variables of touch and verbalization as independent variables. The estimates of b_{1s}, b_{2s} and c_s are the beta weights for the two dimensions and the intercept, respectively. The proportion of variance in preference accounted by the vector model is the R^2 for the overall regression. An F-statistic (with K and $I - K - 1$ degrees of freedom) can be computed to test that the null hypothesis that the R^2 square of the vector model $= 0$ can be computed as follows:

$$F = \frac{R_v^2}{1 - R_v^2} \times \frac{I - K - 1}{K},$$

where K, as before, is the number of dimensions assumed in the model (in this example, $K = 2$), and I is the number of stimuli under investigation (in our example, $I = 11$). Table 6.4 shows the estimated subject weights for each of the dimensions as well as the estimated intercept for each subject.

If a researcher, after having compared the five types of preference models, decided that the vector model did indeed fit the values of the subjects well, the estimates of subject weights may be more parsimoniously used than the entire data on the instrument in describing interindividual differences in performance. The c_s or intercepts of these parameters may also be used to describe differences in overall level of response across the levels investigated.

6.2. *Simple Euclidean Model*

The second model proposed by Carroll, the *simple Euclidean*, assumes that preferences have the following form (except, again, for measurement error):

$$\delta_{is} = \sum_k w_s^2(x_{ik} - x_{sk})^2 + c_s$$

Fitting this model using multiple regression is only slightly more

TABLE 6.4. Salience and Weights Associated with the Vector, Simple Euclidean, Weighted Euclidean, and General Euclidean Models for Four Children

	Touch $W_{\text{touch,verb}}$	Verb.	W	W_{touch}	W_{verb}
Child *A*					
Vector	−.014	.029			
Simple Euclidean		−.06			
Weighted Euclidean			−.10	−.05	
General Euclidean			−.10	−.05	−.03
Child *B*					
Vector	−.025	.060			
Simple Euclidean		−.07			
Weighted Euclidean			−.20	−.05	
General Euclidean			−.10	−.05	.02
Child *C*					
Vector	3.1	.13			
Simple Euclidean			0		
Weighted Euclidean			0	0	
General Euclidean			0	0	.42
Child *D*					
Vector	−.022	0.62			
Simple Euclidean		−.10			
Weighted Euclidean			−.10	−.10	
General Euclidean			−.11	−.10	−.05

complicated than fitting the vector model. For each subject, the dependent variable is still the preferences. The independent variables now, however, are the verbalization and touch variables in addition to a third, constructed independent variable composed of the sum of the squares of each stimulus value.

In addition, it is now possible to decide whether the Euclidean model accounts for a subject's performance or whether the vector model more parsimoniously accounts for it. Table 6.5 shows the statistical tests for the four types of models as well as the overall tests of significance of the model. The table is organized such that, for each subject, the four types of analyses are considered (The weighted Euclidean and general Euclidean models will be presented later.) The first F-statistic in each row

TABLE 6.5. F-Statistics for Overall Significance of Model and comparing the Models against Each Other for Four Children

	r^2	F	Prob(F)		r^2	F	Prob(F)
Child *A*				Child *C*			
Vector model	.19	.96	.42	Vector model	1	4.3×10^{10}	.001
Simple Euclidean model	.83	11.44	.002	Simple Euclidean Model	1	2.5×10^{10}	.001
$F_{\text{Euclid=Vector}}$		26.34	.001	$F_{\text{Euclid=Vector}}$		.04	.956
Weighted Euclidean model	.91	15.88	.002	Weighted Euclidean model	1	1.6×10^{10}	.001
$F_{\text{WEuclid=Vector}}$		25.04	.002	$F_{\text{WEuclid=Vector}}$		.04	.956
$F_{\text{WEuclid=Euclid}}$		5.78	.053	$F_{\text{WEuclid=Euclid}}$		.0	1.00
General Euclidean model	.92	10.89	.010	General Euclidean model	1	1.1×10^{10}	.001
$F_{\text{GEuclid=Vector}}$		14.22	.002	$F_{\text{GEuclid=Vector}}$		.03	.999
$F_{\text{GEuclid=Euclid}}$		2.54	.174	$F_{\text{GEuclid=Euclid}}$		0	1.00
$F_{\text{GEuclid=WEuclid}}$		.13	.731	$F_{\text{GEuclid=WEuclid}}$		0	1.00
Child *B*				Child *D*			
Vector model	.27	1.50	.28	Vector model	.27	1.50	.28
Simple Euclidean model	.67	4.65	.043	Simple Euclidean model	.89	18.48	.001
$F_{\text{Euclid=Vector}}$		8.25	.023	$F_{\text{Euclid=Vector}}$		38.39	.001
Weighted Euclidean model	.96	40.30	.001	Weighted Euclidean model	.89	11.89	.005
$F_{\text{WEuclid=Vector}}$		57.79	.001	$F_{\text{WEuclid=Vector}}$		16.46	.003
$F_{\text{WEuclid=Euclid}}$		49.82	.001	$F_{\text{WEuclid=Euclid}}$		.00	.971
General Euclidean model	.96	27.60	.001	General Euclidean model	.89	8.16	.019
$F_{\text{GEuclid=Vector}}$		33.00	.001	$f_{\text{GEuclid=Vector}}$		9.43	.008
$F_{\text{GEuclid=Euclid}}$		21.37	.003	$F_{\text{GEuclid=Euclid}}$		.07	.937
$F_{\text{GEuclid=WEuclid}}$		.13	.731	$F_{\text{GEuclid=WEuclid}}$		.13	.731

tests the null hypothesis $R^2 = 0$ for that model. For the Euclidean model, it is possible to test the hypothesis that $R_v^2 = R_e^2$ by means of appropriate F-statistics. For the Euclidean model, it is possible to test the hypothesis that $R_v^2 = R_e^2$ by means of appropriate F-statistics. For K dimensions and I stimuli as defined for the vector model, the F-statistic with $K + 1$ and $I - K - 2$ degrees of freedom for testing the hypothesis that $R_e^2 = 0$ is

$$F = \frac{R_e^2}{1 - R_e^2} \times \frac{I - K - 2}{K + 1}$$

The F-statistic with 1 and $I - K - 2$ degrees of freedom to test the hypothesis of $R_e^2 = R_v^2$ against the alternative $R_e^2 > R_v^2$ is

$$F = \frac{R_e^2 - R_e^2}{1 - R_e^2}(I - K - 2)$$

These F-values are shown for the four children in the two rightmost columns of Table 6.5. From this we can see that the vector model fits the data of child C well, since the proportions are perfectly explained in the data, and the F for testing $R_e^2 = R_v^2$ is not significant ($p = .11$). Children A and D, however, are well fit by the Euclidean model. The overall F's for the simple Euclidean model for these children are significant (11.43, 18.48), and the F to test $R_e^2 = R_v^2$ is significant (26.34) and 38.39, respectively). The tests for comparing weighted Euclidean models with the lower-order models fail to achieve statistical significance for these children ($p = .05$ and .97, respectively). Although the F-statistics for child B are significant for the simple Euclidean model, the weighted Eulidean model seems to fit the data best.

The simple Euclidean model is the first preference test that provides a testable hypothesis of the assumption that children's response patterns can be explained in a simple linear fashion. One possible problem in interpretation of these coefficients across several individuals lies in the comparison of several F-tests within a given study. The researcher may wish to use an adjustment to experimentwise alpha to ensure an overall Type I error rate of inference.

One practical interpretation of the simple Euclidean model is that it provides estimates of the ideal level of the two dimensions that would maximize the odds that the child would perform. The farther away a situation is from these points of touch and verbalization, the less the likelihood that the child will comply. The estimated ideal point X_{sk} for subject s on dimension K is equal to

$$x_{sk} = \frac{b_{ks}}{-2w^2},$$

where w^2 is the beta weight of the variable composed of the sums of squares across the dimensions. Table 6.6 shows the estimated ideal points for each of the four subjects and the weight estimates, w^2. For children A and D for example, the ideal point estimates under the simple Euclidean model are (2.0, 2.8) and (2.0, 2.9) for the touch and verbalization dimensions, respectively. This means that both children are most likely to respond to touch of the hand and a complete sentence request to put the toys away. If one assumes, though, that a Euclidean model should be adopted (as might be the case if the researcher concludes that, across all individuals, the Euclidean model provides the simplest explanation of

TABLE 6.6. Ideal Point Estimates for Four Children under the Euclidean, Weighted Euclidean, and General Euclidean Models

	Ideal point touch	Ideal point verbalization
Child *A*		
Simple Euclidean	2.01	2.84
Weighted Euclidean	2.01	2.91
General Euclidean	2.01	2.91
Child *B*		
Simple Euclidean	1.96	2.19
Weighted Euclidean	1.99	2.07
General Euclidean	1.98	2.91
Child *C*		
Simple Euclidean	-3.0×10^{16}	-3.0×10^{16}
Weighted Euclidean	-1.0×10^{16}	-1.0×10^{16}
General Euclidean	-5.9×10^{16}	-6.4×10^{16}
Child *D*		
Simple Euclidean	2.02	2.91
Weighted Euclidean	2.02	2.91
General Euclidean	2.02	2.91

behavior) the ideal point of child C lies beyond the values of the stimulus in the study. This finding is an interesting example of the mathematical relationship of the vector and Euclidean models, namely that the vector model is a simple Euclidean model with ideal point estimates set at infinity.

6.3. The Weighted Euclidean Model

In the simple Euclidean model, the assumption was made that divisions in both dimensions were equally important in determining a subject's preference. No provision was made, for example, that the dimension of verbalization would be more important to a given child than the dimension of touch. In this example, it seems quite reasonable to assume that, since the divisions into categories of touch and verbalization are quite different, an analysis that assumes that the dimensions receive unequal weighting would make more sense.

For the *weighted Euclidean model,* the preferences, in the absence of measurement error, have the form

$$\delta_{is} = \sum_k w_{ks}^2 (x_{ik} - x_{sk})^2 + c_s.$$

This model is, then, identical to the simple Euclidean model, except that separate weights are computed for each dimension. If these weights were the same across all dimensions, the weighted Euclidean model would be equal to the simple Euclidean model. Just as the simple Euclidean model included the vector model as a special case, then so also the weighted Euclidean model includes the simple Euclidean model as a special case.

The F-statistic to test $R_w^2 = 0$ has $2K$ and $I - 2K - 1$ degrees of freedom. Two incremental F-statistics to test $R_w^2 = R_e^2$ and $R_w^2 = R_v^2$ and have $(K - 1, I - 2K - 1)$ and $(K, I - 2K - 1)$ degrees of freedom, respectively. In this example, child C is fit best by the weighted Euclidean model, since the F-statistics comparing the weighted Euclidean model with the simple and vector models are significant. This means that the ideal points for this child are a moderate degree of touch and verbalization in sentences, but departures from the ideal point for verbalization affect the odds of compliance much more than departures from the ideal level of touch, as evidenced by the differences in the weight estimates of .19 and .05 for the verbalization and touch dimensions.

6.4. The General Euclidean Model

The general Euclidean model is an extension of Tucker's (1972) three-mode factor analysis to individual differences scaling. Davison's (1983) matrix presentation is an easy way to understand the model. Consider a diagonal matrix W_s of weights, a transformation matrix T_s for each subject, and a global matrix X, which is a matrix of stimulus by dimensions scaling values. Each subject's unique representation of the stimulus space can then be given by the matrix notation equation

$$X_s = XW_sT_s.$$

Preferences have the following form (assuming no measurement error):

$$\delta_{is} = \sum_k (x_{iks} - x_{sks})^2 + c_s$$

In this formulation of the δ, a subject's dissimilarities are phrased in terms of the subject's idiosyncratic ideal point estimates. To present a form that shows how the δ can be viewed in terms of the matrix formulation given, the δ can also be expressed as

$$\delta_{is} = \sum_k w_{ks}^2 (x_{ik} - x_{sk})^2 + \sum_{kk'} w_{ks}w_{k's}(x_{ik} - x_{sk})(x_{ik'} - x_{sk'}) + c_2,$$

where k is, as before, the dimension of the stimulus space.

What this means to the developmentalist is that the underlying dimensions of the children are assumed to be the same, but that the salient dimensions to the children are within a rotation of each other. Since developmentalists frequently talk not in terms of rotated frames of reference but of issues of increasing dimensionality attended to by the children, we will not take this opportunity to explain all of the F-statistics and incremental F-statistics for testing this model. A full explanation of how to compute these F-statistics is given in Davison (1983). In our example, the general Euclidean model did not represent the response patterns of any of the children. Any of the significant F-statistics comparing the general Euclidean model could be accounted for by the significance of the preceding simpler model(s).

When we consider the implications of the external analysis for developmental psychology, several advantages come readily to mind. First, this model is a workable articulation that can contrast to the models advanced by the traditional "physics" model of behavior, which assumes that all relationships between variables and outcomes act as "forces" and must therefore be linear in nature (Sinnott, 1981). The fact that the vector model is such a model of strictly linear effects provides a test of whether the behaviors of interest can be forced onto such a model. To the extent that the models of probabilities depart from this linear relationship, the developmental model more adequately describes the behaviors at hand. Second, the model provides a readily interpretable set of values that describe the optimal levels of response for a given individual. This information can provide practical information on what treatments may be most appropriate for a given child. Third, as noted in the internal analysis, the stress values (in this case the F-values) associated with a given solution can provide the researcher with some information as to the extent that the individual fits into the external pattern of the underlying dimensions as specified by the researcher. Also, the researcher interested in assessing the degree of fit to the model across a number of individuals could incorporate the numerator sums of squares for a number of individuals in a group and compare this with the performance of a number of other individuals to form a folded F-statistic. The researcher would be making the assumption that the individuals in the numerator and the denominator were interchangeable, but this might be justifiable if one has reason to believe in the interchangeability of individuals across a given class.

6.5. Overall Conclusions

Nonmetric methods can, as indicated in the first example, be used to recover the underlying relationships between variables in terms of an overall geometric representation. In the presence of a strong theory about the underlying dimensions, nonmetric multidimensional scaling methods can also be used to test hypotheses about the underlying optimal levels of the dimensions that the individual seeks to maximize in interaction with the environment. Notice that the estimation of these points was done with a much smaller sample size than that required by techniques usually employed in developmental psychology. To conduct a similar study using P-technique, a much larger series of samples would have had to have been drawn in order to make conclusions about the individual. The F-statistics that have been presented to compare and contrast the vector, Euclidean, and weighted Euclidean models allow for different underlying measurement models to be tested at the individual level. Nonmetric techniques could be used, for example, as a parsimonious summary of the characteristics of an individual. The ideal points, for example, from several individuals along the external dimensions presented in the second example could summarize several variables' performance for an individual into one or two variables.

Some of the other advantages of nonmetric scaling in developmental research have been explored in methodological papers that seek to describe how the obtained solutions of nonmetric scaling relate to more traditional measures, such as principal components factor analysis. Davison (1985) found, in several studies, that the factor loadings for 2 through k factors were for most practical purposes, equal to the scaling solution in the 1 through $k - 1$ dimensions. As he pointed out, in analysis of dissimilarity data, the nonmetric solution proceeds based on the double-centered matrix of dissimilarities. This means that the nonmetric solution is based on the overall level of differences between one individual and another subtracted out of the responses. As such, then, if there exist large interindividual differences between individuals, nonmetric scaling will subtract these out before examining the relationships between stimuli. There are many cases in assessment of attitudes when it might be desirable to perform this correction. Davison and Srichantra (1988) examine components analysis solutions and nonmetric scaling solutions when a large acquiescence factor is present in the data. In these cases, the acquiesence factor is to be largely ignored in the interpretation of the solution, and the researcher is interested in the other relationships among the variables in the study.

9. Computational Issues

In the practical use of nonmetric analyses in summarizing data, there are three problems commonly encountered: local minima, degenerate solutions, and lack of convergence.

9.1. Local Minima

As explained, nonmetric scaling programs are iterative and seek to minimize some measure of stress. A local minimum occurs when the algorithm, in attempt to compute optimal values for distances between stimuli, falsely determines that a given distance model is the solution with the lowest stress value when in fact it is not. The program, when running with a different starting configuration, may arrive at another minimum value under the algorithm, which may or may not be the "true" global minimum for the stress function. Three things can be done to help avoid the problem. First, it has been this author's experience that local minima are particularly a problem when using many of the older scaling programs. Newer programs seem to avoid this problem by using more sophisticated iterative approaches. Second, problems of local minima often occur due to using an inappropriate stress formula, given the data to be analyzed. It was noted that use of stress formula 1 leads to fewer computational difficulties when the input data are symmetric dissimilarities. When the data are preferences, a situation that commonly results in the data being unsymmetric rectangular matrices, then formula 2 leads to fewer problems. Still, even when newer computer implementations are employed and appropriate stress values are used, local minima can occur. The most common technique to detect and/or get around this problem is to employ the iterative algorithm for estimating scale values with various initial values of stress coefficients. What is usually done in nonmetric computer programs is to compute the initial, metric values for the stimulus coordinates and then to pass these initial metric values to the iterative portion of the program. In order to start the nonmetric phase of the program at different locations on the stress function to be minimized, different starting values for the stimulus space can be used. Hopefully, use of several different starting values will result in finding stress values that constitute a global minimum over the stress function.

7.2. Degenerate Solutions

A degenerate solution occurs in nonmetric scaling when the number of estimated stimulus values is small relative to the number of stimulus points. For example, if stimulus points G_{12}, G_{22}, and G_{32} all received the same estimated stimulus values in the scaling solution shown in Figure 6.2, this would be an example of a degenerate solution. The only way to determine if a degenerate solution has occurred is by inspection of the estimated stimulus points. Davison (1983) warns that degenerate solutions occur particularly in one and two dimensions. Shepard (1974) gave the rule of thumb that when stress values near zero are obtained users should be particularly careful to inspect the solutions for degeneracies.

7.3. Lack of Convergence

Lack of convergence occurs when the iterative portion of the nonmetric solution fails to achieve a solution within the iterations allowed by the program. When this occurs, one of the easiest strategies to solving this is to increase the maximum number of iterations allowed by the program. In situations where the user is using an older program or where the solution is particularly difficult to compute, the obtained solution may be significantly affected by rounding error induced during the execution of the program. In these cases, the user may wish to rerun the solution so obtained, but input the end configuration as the starting configuration.

Although these three computational problems may sound difficult, they rarely occur in practice. In general, they can be avoided by using the newer implementations of the algorithms, and in some situations can be remedied by using several start values.

References

Arnold, J. B. (1971). "A multidimensional scaling study of semantic distance," *Journal of Experimental Psychology Monograph.* **90,** 349–372.

Attneave, F. (1950). "Dimensions of similarity," *American Journal of Pyschology.* **63,** 516–556.

Bechtel, G. G. (1976). *Multidimensional preference scaling.* The Hague, Netherlands: Mouton.

Bechtel, G. G., Tucker, L. R., and Chang, W. A. (1971). "A scalar product model for multidimensional scaling of choice," *Psychometrika.* **36,** 369–388.

Carroll, J. D. (1972). "Individual differences and multidimensional scaling." In R. N. Shepard, A. K. Romney, and S. B. Nerlove (Eds.), *Multidimensional scaling: theory and applications in the behavioral sciences*; Vol. 1. (pp. 105–153) New York: Seminar Press.

Coombs, C. H., McClelland, G. H., and Coombs, L. C. (1973). *The measurement and analysis of family composition preferences.* Michigan Mathematical Psychology Program, MMPP 73–5.

Davison, M. L. (1983). *Multidimensional scaling.* New York: Wiley.

Davison, M. L. (1985). "Multidimensional scaling versus components analysis of test intercorrelations," *Psychological Bulletin.* **97,** 94–105.

Davison, M. L., Richards, P. S., and Rounds, J. B. (1986). "Multidimensional scaling in counselling research and practice," *Journal of Counseling and Development.* **65,** 178–184.

Davison, M. L., and Srichantra, N. (1988). "Acquiescence in components analysis and multidimensional scaling of self-rating items," *Applied Psychological Measurement.* **12,** 339–351.

Davison, M. L., and Wood, P. K. (1983). "Fitting unidimensional choice models with nonmetric multidimensional scaling," *Applied Psychological Measurement.* **7**(3), 333–340.

Feinberg, S. E. (1980). *The analysis of cross-classified categorical data,* 2nd ed. Cambridge, Mass.: MIT.

Ford, D. (1987). *Humans as self-constructing living systems, a developmental perspective on behavior and personality.* Hillsdale, N.J.: Lawrence Erlbaum.

Guttman, L. (1968). "A general nonmetric technique for finding the smallest coordinate space for a configuration of points." *Psychometrika,* **33,** 469–506.

Kruskal, J. B. (1964). "Multidimensional scaling by optimizing goodness-of-fit to a nonmetric hypothesis," *Psychometrika.* **29,** 1–28, 115–129.

Kruskal, J. B., and Carroll, J. D. (1969). "Geometric models and badness-of-fit functions." In P. R. Krishnaiah (Ed.), *Multivariate analysis,* Vol. 2. New York: Academic Press.

Labouvie, E. W. and Nesselroade, J. R. (1989). Structural modeling for longitudinal data: Some conceptual and empirical explorations available from the Center for Alcohol Studies, Rutgers University, New Brunswick, N.J.

Lingoes, J. C. (1973). *The Guttman-Lingoes nonmetric program series.* Ann Arbor, Mich.: Mathesis Press.

Nesselroadle, J. R., and Cattell, R. B. (1989). *Handbook of multivariate experimental psychology.* New York: Plenum Press.

Nesselroade, J. R., and Ford, D. (1987). "Putting the framework to work: methodological implications of the systems framework." In D. Ford (Ed.), *Humans as self-constructing living systems: a developmental perspective on behavior and personality* (pp. 47–79). Hillsdale, N. J.: Lawrence Erlbaum.

Ramsey, J. O. (1986). *Multiscale II manual.* Montreal, Canada: Department of Psychology, McGill University.

Richardson, M. W. (1938). "Multidimensional psychophysics," *Psychological Bulletin.* **35,** 659–660.

Rosenberg, S., and Jones, R. (1972). "A method of investigating and representing a person's implicit theory of personality: Theodore Dreiser's view of people," *Journal of Personality and Social Psychology.* **22,** 372–386.

Roskam, E. E. (1969). Data theory and algorithms for nonmetric scaling, parts 1 and 2. Unpublished manuscript. Catholic University, Nihmegan, The Netherlands.

Shepard, R. N. (1962). "The analysis of proximities: multidimensional scaling with an unknown distance function," *Psychometrika.* **27,** 125–140, 219–246.

Shepard, R. N. (1972). "Psychological representation of speech sounds." In E. E. David and P. B. Denes (Eds.), *Human communication: a unified view* (pp. 67–113). New York: McGraw-Hill.

Shepard, R. N. (1974). "Representation of structure in similarity data: problems and prospects," *Psychometrika.* **39,** 373–422.

Sherman, C. R. (1972). "Nonmetric multidimensional scaling: A Monte Carlo study of the basic parameters," *Psychometrika.* **37,** 323–355.

Sinnott, J. D. (1981). "The theory of relativity: a methatheory for development?" *Human Development.* **24,** 293–311.

Takane, Y., Young, F. W., and DeLeeuw, J. (1977). "Nonmetric individual differences multidimensional scaling: an alternative least squares method with optimal scaling features," *Psychometrika.* **42,** 7–67.

Torgerson, W. S. (1952). "Multidimensional scaling. I. Theory and method," *Psychometrika.* **17,** 401–419.

Tucker, L. R. (1960). "Intra-individual and inter-individual multidimensionality." In H. Gulliksen and S. Messick (Eds.), *Psychological scaling: theory and applications.* New York: Wiley.

Tucker, L. R. (1972). "Relations between multidimensional scaling and three-mode factor analysis," *Psychometrika.* **37,** 3–28.

von Eye, A., and Kreppner, K. (1989). "Family systems and family development: the selection of analytical units." In K. Kreppner and R. M. Lerner (Eds.), *Family systems and life-span development.* Hillsdale, N.J.: Lawrence Erlbaum.

Wiener-Ehrlich, W. K. (1978). "Dimensionality and metric structures in multidimensional stimuli," *Perception and Psychophysics.* **24,** 399–414.

Young, F. W. (1987). *Multidimensional scaling—history, theory and applications.* (Robert Hamer, Ed.) Hillsdale, N.J.: Lawrence Erlbaum.

Young, F. W., and Lewyckyj, R. (1979). *ALSCAL 4 user's guide,* 2nd ed. Chapel Hill, N.C.: Data Analysis and Theory Associates.

Author Index

Numbers in italic indicate pages on which complete references are given

K

L

M

Subject Index

Roman page numbers refer to Volume I; italic page numbers refer to Volume II.

T

STATISTICAL MODELING AND DECISION SCIENCE

Gerald J. Lieberman and Ingram Olkin, editors

Samuel Eilon, *The Art of Reckoning: Anaysis of Performance Criteria*
Enrique Castillo, *Extreme Value Theory in Engineering*
Joseph L. Gastwirth, *Statistical Reasoning in Law and Public Policy: Volume 1, Statistical Concepts and Issues of Fairness; Volume 2, Tort Law, Evidence, and Health*
Takeaki Kariya and Bimal K. Sinha, *Robustness of Statistical Tests*
Marvin Gruber, *Regression Estimators: A Comparative Study*
Alexander von Eye, editor, *Statistical Methods in Longitudinal Research: Volume I, Principles and Structuring Change; Volume II, Time Series and Categorical Longitudinal Data*